SMALL
GAS
ENGINES

SMALL GAS ENGINES

Second Edition

James A. Gray
Richard W. Barrow

School of Technology
Indiana State University

Prentice Hall
Englewood Cliffs, New Jersey 07632

Library of Congress Cataloging-in-Publication Data

Gray, James A., 1937–
  Small gas engines.

  Includes index.
  1. Internal combustion engines—Maintenance and
repair. I. Barrow, Richard W. II. Title.
TJ789.G66 1988   621.43 ′4   87–25939
ISBN 0-13-813171-6
ISBN 0-13-813189-9 (pbk.)

Editorial/production supervision and
  interior design: Ed Jones
Cover design: George Cornell
Manufacturing buyers: Lorraine Fumoso, Peter Havens
Page layout: Lisa Botto

Drawings by *Robert F. MacFarlane*

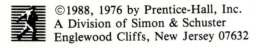
Printed in the United States of America

10 9 8 7 6 5 4 3 2 1

ISBN   0-13-813171-6

ISBN   0-13-813189-9 {PBK}

PRENTICE-HALL INTERNATIONAL (UK) LIMITED, *London*
PRENTICE-HALL OF AUSTRALIA PTY. LIMITED, *Sydney*
PRENTICE-HALL CANADA INC., *Toronto*
PRENTICE-HALL HISPANOAMERICANA, S.A., *Mexico*
PRENTICE-HALL OF INDIA PRIVATE LIMITED, *New Delhi*
PRENTICE-HALL OF JAPAN, INC., *Tokyo*
SIMON & SCHUSTER ASIA PTE. LTD., *Singapore*
EDITORA PRENTICE-HALL DO BRASIL, LTDA., *Rio de Janeiro*

CONTENTS

Note: For coverage of specific subjects, see
index, pages 255-257.

PREFACE

The intent of this manual is to provide the individual with enough information and guidance so that he or she can become actively involved with engines. The manual explains and illustrates the basic operational theory and service procedures for single-cylinder air-cooled engines. The content is organized so that it can be used as a guide in diagnosing as well as performing minor and major engine service.

This manual is a resource for individuals actively engaged in the maintenance and repair of the small engine and a text or reference for academic instruction and reference. Teachers who wish to make transparency masters from the illustrations may do so without securing permission from the publisher.

Numerous tips and checks are presented that help bridge the gap between theory and actual practice. Handbooks containing specific instructions and specifications are available from the engine manufacturers. These are an invaluable aid to the person servicing the small gas engine.

This book presents the "why" and the "how to" in individual packages with illustrations to tell the story. In-depth scientific theory is *minimized* in favor of a fundamental understanding of how the theory works.

The manual also is intended for the home mechanic who performs his or her own maintenance and service on the small-engine-powered, labor-saving equipment common today.

Appreciation is expressed to the wives of the authors, Phyllis A. Gray and Virginia L. Barrow, who provided invaluable assistance in the preparation and typing of the manuscript.

James A. Gray
Richard W. Barrow

Section I

SMALL GAS ENGINE QUICK REFERENCE: SAFETY PRECAUTIONS, MAINTENANCE, TROUBLESHOOTING, AND STORAGE

SAFETY PRECAUTIONS

Always think through any operation before starting to work to be certain you are observing all safety precautions.

Read the operator's manual before starting the engine or using powered equipment.

Don't store gasoline-powered equipment in a garage or storeroom in which there is a furnace, water heater, or any other gas-fired or open flame-type unit.

Wear safety glasses when operating equipment, adding fuel, or performing adjustments or service operations. Remove jewelry (rings and watches) especially when working with electrical components. Do not make checks or adjustments on running engines or equipment unless it is absolutely necessary. BE EXTREMELY CAUTIOUS of drive mechanisms, belts, and blades or cutters. DISENGAGE POWER before performing any checks or services to the drive train.

NEVER add fuel to an engine inside a closed building. Always move unit outdoors before adding fuel. Likewise, always mix fuel for two-cycle engines outdoors.

Do not permit small children to operate or be in close proximity to powered equipment.

Thoroughly explain and teach the safety precautions to any inexperienced person who is planning to use your equipment.

Ground the spark plug wire before doing any service operations on the driven blade or unit.

Avoid storing equipment that is extremely hot. Allow unit to cool before storing it indoors. Keep the engine clean, especially the area around the exhaust system.

Do not permit sparks, fire, or flame near the fuel tank, carburetor, battery, or fuel storage can. NEVER SMOKE while performing service operations on or near the engine battery or fuel storage.

Observe safety precautions when cleaning or servicing the battery. Rinse the battery hydrometer thoroughly after each use.

Dispose of used crankcase oil in an approved manner. DO NOT pour it into a water drain or sewage system.

Discard oily rags immediately. Do not permit oily rags to accumulate in a pile as they may ignite from spontaneous combustion.

Do not permanently remove shields or guards from equipment. ALWAYS replace all housings, shields, and guards upon completion of servicing operations.

Do not disconnect or disable "interlock" electrical safety mechanisms or devices.

On rope-start equipment, observe safety precautions when placing your feet before pulling the starter rope.

QUICK REFERENCE TO ENGINE MAINTENANCE

SMALL GAS ENGINE MAINTENANCE SCHEDULE
FOR TYPICAL HOMEOWNER USE

Unit		Each use	Each season or 25 hrs., whichever comes first	Every 6 mos. or 100 hrs., whichever comes first	Each year or 300 hrs.	Reference page
Engine oil (4-cycle only)	Check	x				47
	Change		x*			47
Spark plug	Clean & adjust			x		6, 57
	Replace				x	6, 57
Air cleaner	Check	x				105
	Clean or replace		x*			105
Fuel filter or strainer	Clean or replace				x	95
Cooling fins and screens	Check & clean as needed	x				49
	Remove shrouds & clean				x*	49
Combustion chamber	Remove carbon				x	119
Carburetor	Adjust			x		43
	Clean externally			x		103
Ignition points & condenser (if so equipped)	Replace				x	71
Battery	Check fluid level (where applicable)		x			203
	Clean connections				x	203
Drive belts	Check or adjust		x			221

*Service more frequently when used under extremely dirty conditions.

5

SPARK PLUG DIAGNOSIS

CONDITION	POSSIBLE CAUSE	CORRECTION
Normal	Gray or light tan deposits on insulators; electrodes— not burned away; gap at specifications	
Gap bridged with carbon	Excess deposits in combustion chamber	Clean or replace plug Clean combustion chamber
Preignition or melted electrodes	Engine overheating	Check cooling system
	Wrong type fuel	Replace fuel
	Over-advanced ignition timing	Reset timing
	Wrong type spark plug	Replace spark plug
	Lean fuel mixture	Correct air intake leaks or readjust carburetor
"Fluffy" deposits	Wrong type fuel	Replace with different fuel
Cold fouling or excessive carbon deposits	Over-rich fuel mixture	Check and adjust carburetor
	Clogged air filter	Replace air filter
	Sticking valves or worn valve guides	Inspect valve train
Wet fouling	No "spark"	Check ignition system and correct as needed
	Flooded engine	Check choke and carburetor adjustments and correct as needed
Wet fouling (Oil deposits)	Worn rings and piston	Overhaul engine as needed
	Worn or loose bearings	Correct as needed

GENERAL DATA
AND SPECIFICATIONS

You should adhere to the recommendations and specifications provided by the engine manufacturer. Follow the information and specifications listed below *only* in instances where the manufacturer's data are not available.

4-STROKE CYCLE ENGINE

FUEL	Regular leaded or regular unleaded
OIL	Always use a good grade (A.P.I. rating SC, SD, SE, or SF)
	Weight Summer (above 40°F) SAE 30 or 10W30 Winter (below 40°F) SAE 10W or 10W30 Winter (below 0°F) SAE 10W
	Use of 10W40 oil is **not** *recommended*
	Fill crankcase until oil level shows on full mark on dipstick or until oil is within 1/8″ (3mm) of top of crankcase filler plug
SPARK PLUG	*Type* Replace with same type as was removed or refer to application chart at parts store
	Gap .030″ (.76 mm)
	Torque "Snug" plus 1/3 turn or 15–20 ft. lb. (20–27 NM)
IGNITION POINT GAP	.020″ (.51 mm)
ARMATURE AIR GAP	*Two leg* .008″–.010″ (.20–.25 mm)
	Three leg .014″–.022″ (.36–.60 mm)
IDLE R.P.M.	1300–1800

2-STROKE CYCLE ENGINE

FUEL	Regular leaded or regular unleaded mixed with 2-cycle engine oil. Mix as specified on the engine decal. If mix ratio is not listed, mix 1/2 pint of 2-cycle oil with 1 gallon gasoline. Nondetergent oil designated API, SA, or SE, SAE 40 may be substituted for 2-cycle oil. *Do not substitute* multiple viscosity oils (10W30) or detergent oils.
SPARK PLUG	Replace with same as was removed or refer to chart at parts store
	Gap .035″ (.88 mm)
	Torque 12–15 ft. lb. (16–20 NM) or snug plus 1/3 turn
IGNITION POINT GAP	.020″ (.51 mm)
ARMATURE AIR GAP	.010″ (.25 mm)

TROUBLESHOOTING
AN "ILL" ENGINE

CONDITION: ELECTRIC START ENGINE WON'T CRANK

POSSIBLE CAUSES	CORRECTION	REFERENCE PAGE
Unit not in "neutral"	Place unit in neutral and disengage the drive train (operator must be on seat and clutch must be depressed on some equipment).	
Loose cable connection	Clean and tighten both battery cable connections	203
	Tighten cable connections at ground (on unit frame), at the starter solenoid, and at the starter	
Weak or dead battery	Test battery and recharge or replace as needed	205
Loose starter drive belt	Inspect belt and tighten or replace as needed	221
Loose connections on ignition switch	Check connections on switch and tighten as needed	217
Defective solenoid	Test solenoid and replace if faulty	217
Defective starter	Test starter and replace or repair as needed	221, 223

CONDITION: ENGINE CRANKS BUT WON'T START

POSSIBLE CAUSES	CORRECTION	REFERENCE PAGE
No fuel in tank	Refill tank with fresh fuel	
Unit in gear	Be certain unit is in neutral and all safety requirements for starting are observed	
Fuel line shut off	Open shut-off valve (if so equipped)	101
Check oil level	Add oil to proper level	47
Defective spark plug	Remove spark plug, clean, and regap or replace with new spark plug.	6, 57
Faulty ignition (no spark at spark plug)	Check for "grounds" of ignition wires: Test ignition system components. Check ignition points—condition and adjustment. Check for worn main bearing on crankshaft, which will cause improper point contact.	59–81

TROUBLESHOOTING
AN "ILL" ENGINE

CONDITION: **ENGINE CRANKS BUT WON'T START** (*cont.*)

POSSIBLE CAUSES	CORRECTION	REFERENCE PAGE
Engine flooded (excess fuel in cylinder)	Remove spark plug: If electrode area of plug is wet with fuel, allow fuel to evaporate: Pour one teaspoon of SAE 30 oil into cylinder and replace spark plug. If spark plug electrode area is dry, engine may not be getting enough fuel.	
Insufficient fuel:		
insufficient choking	Apply full choke and attempt to start.	87
tank vent clogged	Open vent in tank cap or clean cap threads which provide venting	101
clogged fuel line	Disconnect line and clean	
dirty fuel filter	Replace filter	
water in fuel or stale fuel	Drain fuel and refill tank with fresh fuel	
faulty fuel pump (if so equipped)	Replace pump diaphragm or pump unit as required	95–97
Poor compression (engine cranks too freely)	Check compression. If low, check for valve stuck open, leaking head gasket, or worn rings. (See testing section.) Correct problem as noted in engine overhaul section.	39
Improperly adjusted carburetor	Readjust carburetor	43
Clogged air cleaner	Clean or replace air cleaner	105

9

TROUBLESHOOTING
AN "ILL" ENGINE (cont.)

CONDITION: ENGINE IS HARD TO START

POSSIBLE CAUSES	CORRECTION	REFERENCE PAGE
Lean fuel mixture	Check and correct loose manifold or carburetor	101, 177
	Be certain choke is operating	
Carburetor out of adjustment	Check and/or correct carburetor adjustments	43
Faulty spark plug	Replace spark plug	6, 57
Fuel filter clogged	Replace fuel filter	
Fuel line clogged	Clean fuel line	
Faulty fuel pump unit (if so equipped)	Replace pump diaphragm or pump unit as required	95–97
Wrong type fuel or water in fuel	Replace with "good" fuel	
Engine overchoked	Remove spark plug, allow fuel to evaporate, and replace spark plug	105
	Clean and replace air cleaner	
Ignition timing off	Adjust timing	75
Points out of adjustment (if equipped)	Adjust or replace points	
Defective ignition condenser	Replace condenser	71
Defective ignition coil or ignition components	Test and replace as needed	73
Governor out of adjustment	Adjust as needed	99
Poor compression	Replace head gasket or perform major engine service as needed	119

TROUBLESHOOTING
AN "ILL" ENGINE (cont.)

CONDITION: ENGINE LACKS POWER

POSSIBLE CAUSES	CORRECTION	REFERENCE PAGE
Choke stuck "on" or carburetor out of adjustment	Adjust as needed 43	
Faulty spark plug	Replace spark plug 6, 57	
Partially sheared flywheel key	Replace key...................... 67	
Ignition timing off	Adjust as needed 75	
Defective ignition coil or condenser	Test and replace as needed 73	
Low on oil (improper oil mix for 2-cycle engine)	Add oil as needed	
Weak compression due to:		
Sticky valves	Clean valves and guides .... 149 Replace valve springs	
Burned valves	Service valve train 151–153	
Worn rings/ piston	Overhaul engine 114	
Restricted exhaust system	Clean exhaust ports (2-cycle engines only)..................... 177 Replace faulty muffler or exhaust pipe 167	
Restricted air intake	Clean or replace air cleaner........ 105	
Lean mixture due to: Loose carburetor	Check gasket (replace if defective) and tighten carburetor to manifold	
Incorrect mixture	Adjust carburetor 43	

CONDITION: ENGINE MISFIRES WHEN UNDER LOAD

POSSIBLE CAUSES	CORRECTION	REFERENCE PAGE
Faulty spark plug	Remove spark plug, clean, and regap or replace with new spark plug 6, 57	
Improper carburetor adjustment	Readjust carburetor	
Loose ignition wire	Check and tighten ignition wire connections 43	

TROUBLESHOOTING
AN "ILL" ENGINE (cont.)

CONDITION: ENGINE LOSES POWER— MISFIRES OR STOPS AFTER WARMED UP

POSSIBLE CAUSES	CORRECTION	REFERENCE PAGE
Defective condenser	Test and replace as needed	71
Defective ignition coil	Test and replace as needed	73
Defective electronic ignition unit	Test and replace as needed	81
Engine overheating	Clean cooling passages and fins	49
	Check oil level (oil mix for 2-cycle engines)	
Fouled spark plug	Clean and regap or replace plugs	6, 57
Sticking valves	Clean valve guides: Replace valve springs	149

CONDITION: ENGINE RUNS UNEVENLY—SURGES

POSSIBLE CAUSES	CORRECTION	REFERENCE PAGE
Dirty air cleaner	Clean or replace air cleaner	105
Improper carburetor adjustment	Readjust carburetor	43
Defective pump or choke diaphragm in carburetor	Replace diaphragm	95–97
Defective needle valve or float in carburetor	Readjust float level or replace components as needed	95
Intake leak	Tighten intake manifold to engine: Tighten carburetor to intake manifold. (NOTE: Be certain gaskets are not damaged.)	177
Dirty carburetor	Remove carburetor, clean, and install "repair kit."	103
Speed control out of adjustment	Readjust speed control (governor) as needed	
Clogged muffler or exhaust ports (2-cycle engines only)	Clean carbon from muffler and exhaust ports	177

TROUBLESHOOTING
AN "ILL" ENGINE (cont.)

CONDITION: ENGINE RUNS HOT

POSSIBLE CAUSES	CORRECTION	REFERENCE PAGE
Clogged air passages	Clean debris from intake at flywheel, all fins, and air passages. Make certain *all* sheet metal shrouds are in place.	49
Ignition timing off	Check and adjust timing............	75
Lean fuel mixture	Adjust carburetor. Correct vacuum leaks. (Make certain that carburetor and manifold are securely attached.)	
Lack of lubrication	Make sure crankcase oil is at proper level and sufficiently clean. (For 2-cycle engines, make certain oil mixture is correct.)	
Overloaded engine	Reduce load on engine	
Carbon buildup in combustion chamber	Remove cylinder head and clean carbon from chamber. On 2-cycle engines, also clean exhaust ports.	119, 177

CONDITION: ENGINE KNOCKS

POSSIBLE CAUSES	CORRECTION	REFERENCE PAGE
Loose flywheel, blade, or driven unit	Disconnect spark plug wire. Attach to ground, and tighten flywheel or driven unit as required.	
Low on engine oil	Shut off engine and check oil. Add proper oil as needed. (If unit was low on oil and still knocks when restarted, shut unit off as it probably needs major repair.) On 2-cycle engines, check the oil-fuel mixture.	
Carbon buildup in combustion chamber	Remove cylinder head and clean carbon from chamber...............	119, 177
Worn cylinder or collapsed piston	Disassemble engine and recondition as needed	114, 181
Loose connecting rod	Disassemble engine and replace worn components	114, 181

TROUBLESHOOTING
AN "ILL" ENGINE (cont.)

CONDITION: **ENGINE LEAKS OIL FROM BREATHER
(4-CYCLE ENGINE ONLY)**

POSSIBLE CAUSES	CORRECTION	REFERENCE PAGE
Oil level too high	Drain oil from crankcase until level is at "full" mark	
Defective breather	Remove breather assembly, clean, and replace components as needed. Be certain drain hole in chamber is open.	125
Breather improperly assembled	Remove breather assembly and make certain drain holes are down.	125
Breather assembly loose or gaskets leaking	Install new gaskets and tighten securely.	125
Loose oil filler cap or defective seal	Check seal of filler cap to be certain it is not leaking. Replace seals as needed.	
Engine speed too fast	Check governed speed against manufacturer's specifications. Adjust as needed.	
Compression leaking past piston rings into crankcase	Piston ring end gaps aligned, worn rings, worn cylinder; perform compression test—see testing section.	159

TROUBLESHOOTING
AN "ILL" ENGINE (cont.)

CONDITION: **ENGINE USES TOO MUCH OIL**
(4-CYCLE ENGINES ONLY)

POSSIBLE CAUSES	CORRECTION	REFERENCE PAGE
Faulty breather	(See previous section on oil leaking from breather.)	125
Oil leaks	Check for leaks at seals, filler pipes, and crankcase sealing points. Correct as needed.	
Oil level too high (blue smoke)	Check to be certain the proper amount of oil was installed in the engine. Check for defective dipstick.	
Worn valve guides (blue smoke)	Remove cylinder head and check for oily carbon buildup under intake valve head. If buildup is present, remove valves, ream out guides, and install valves with oversized stems.	119
Worn cylinder wall, rings, piston (blue smoke)	Check compression (see testing section). Rebuild engine as needed.	39
Oil passages clogged	Disassemble engine and clean passages as needed.	121

CONDITION: **ENGINE VIBRATES**

POSSIBLE CAUSES	CORRECTION	REFERENCE PAGE
Engine loose on mounting	Tighten attaching bolts as needed	
Bent crankshaft	Check crankshaft	127
Bent or unbalanced driven unit	Check blade or driven unit	

CONDITION: **ELECTRICAL CIRCUIT PROBLEMS**

POSSIBLE CAUSES	REFERENCE PAGE
Ignition	See Section V, page 51
Battery	See Section X, page 201
Starter	See Section X, page 201
Charging	See Section XI, page 227

SMALL GAS ENGINE STORAGE

Figure 1

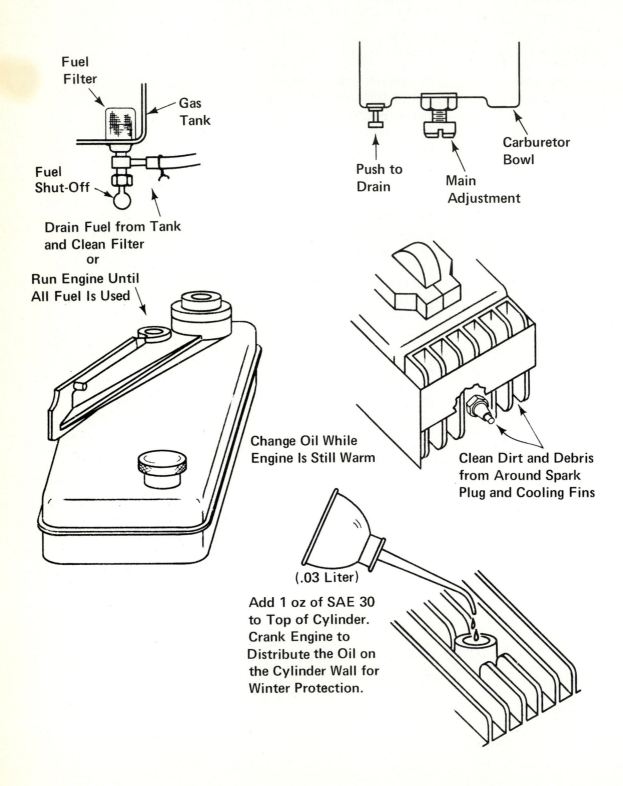

Fuel Filter

Gas Tank

Fuel Shut-Off

Drain Fuel from Tank and Clean Filter
or
Run Engine Until All Fuel Is Used

Push to Drain

Main Adjustment

Carburetor Bowl

Change Oil While Engine Is Still Warm

Clean Dirt and Debris from Around Spark Plug and Cooling Fins

(.03 Liter)

Add 1 oz of SAE 30 to Top of Cylinder. Crank Engine to Distribute the Oil on the Cylinder Wall for Winter Protection.

SMALL GAS ENGINE STORAGE

The successful operation of a small gas engine depends on the care given it. A few precautions taken before storing an engine can prevent several problems that can occur during storage.

REMOVE ALL FUEL

Fuel that evaporates from an engine leaves gummy deposits that can cause serious problems. Drain the fuel tank and clean the filter screen. Drain the carburetor. On some models this can be done by pushing the small drain on the bottom of the carburetor fuel bowl. An excellent way to drain the fuel system is to run the engine until the fuel is used up. This also warms up the engine so that the oil change will remove most of the dirt accumulated in the crankcase.

CHANGE ENGINE OIL (FOUR-STROKE CYCLE ONLY)

Drain engine oil and refill crankcase with new oil as recommended by the manufacturer. See page   if manufacturer data are not available.

CLEAN THE ENGINE

If necessary, remove air shrouds to clean around the cooling fins, governor linkage, and spark plug. Always replace the air shrouds, for they are an essential part of the cooling system.

LUBRICATE THE CYLINDER WALL

Remove the spark plug (see Spark Plug Service). Add approximately one ounce (.03 liter) of SAE 30 motor oil to the cylinder. Make sure that the piston is near top dead center before adding oil, especially on two-stroke cycle engines. Crank the engine several times to distribute the oil all over the cylinder wall area, valves, and valve guides. Replace the spark plug with a new plug (see Spark Plug Service).

REMOVE THE BATTERY

If a unit is battery equipped, the battery should be removed during storage. Care should be taken to record the battery's position and cable connections. If the battery cables are connected backward, the alternator diodes will be blown instantly.

Place the battery in a cool, dry place where it will not *freeze*. The battery should be fully charged when it is placed in storage and it should be recharged occasionally to maintain full charge (see Battery Section). Do not store battery on concrete. Place a small board under the battery during storage to keep it off the concrete.

RESTARTING AFTER STORAGE

1. Check oil level.
2. Clean air filter.
3. Check battery electrolyte level (see Battery Service).
4. Add clean, *fresh* fuel.
5. Adjust carburetor.

Questions for Section I

T F   1. Engine oil should be checked each time the engine is used. Page 5.

T F   2. The spark plug should be replaced every 25 hours of use. Page 5.

T F   3. The fuel filter should be replaced once a year or 300 hours of use. Page 5.

T F   4. The cooling fins and screens should be checked to see if they are clogged each time the engine is used. Page 5.

T F   5. The battery connections should be cleaned at least once a year. Page 5.

   6. How frequently should the air cleaner be serviced? Page 5.

   7. What American Petroleum Institute oil ratings are acceptable for the crankcase oil installed in most small gas engines? Page 7.

   8. What SAE-weight oil is recommended for summer use in the small gas engine? Page 7.

   9. What oils may be used for mixing with the gasoline for most 2-stroke cycle engines? Page 7.

  10. What should one check first when troubleshooting an engine that cranks but won't start? Page 8.

  11. How can a flooded engine be cleared out or corrected? Page 9.

  12. What component is checked first on an engine that misfires when under load? Page 11.

  13. Name three faults that can cause an engine to "surge." Page 12.

  14. State the first thing you should check to correct an engine overheating problem. Page 13.

  15. Identify three faults that can cause an engine to knock. Page 13.

  16. What faults can cause an engine to use excessive oil? Page 15.

  17. Describe the appearance of a "normal" spark plug removed from an engine. Page 6.

  18. Why should the fuel be removed from the fuel tank when the engine is stored? Page 17.

  19. Why should you record or mark the battery's connections when removing it from the unit? Page 17.

Section II

INTRODUCTION TO SINGLE-CYLINDER, AIR-COOLED ENGINES

THE BASIC ENGINE

Figure 2

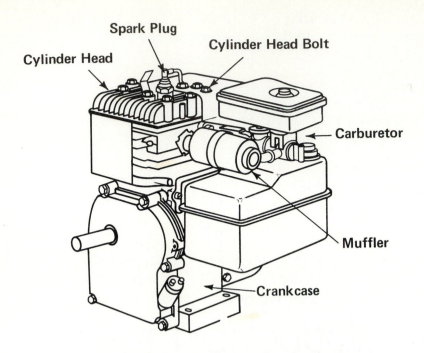

HORIZONTAL CRANKSHAFT ENGINE

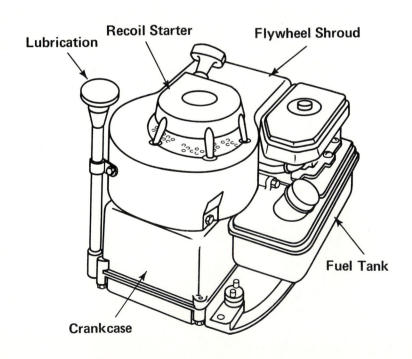

VERTICAL CRANKSHAFT ENGINE

THE BASIC ENGINE

Small engines are classified as being either vertical or horizontal shaft types. This identification refers to the position of the crankshaft of the engine.

Most rotary lawnmowers use a vertical shaft engine.

The small engine is made up of several systems that work together to produce power. The basic systems are as follows:

1. Fuel and carburetion
2. Ignition
3. Lubrication
4. Cooling
5. Exhaust

These systems all work together to convert air and fuel into mechanical energy. Each must serve its own function and be timed properly to the other systems as well as to the piston-crankshaft assemblies.

The FUEL and CARBURETION system provides the proper air-fuel mixture for the engine to operate under all conditions. Proper operation of this system is dependent upon a supply of clean, fresh fuel in the tank and proper adjustments at the carburetor.

The IGNITION system must provide the voltage necessary to jump the spark plug gap at exactly the right time. For this system to do its job, the spark plug, wires, and all other electrical components must be in good condition and properly adjusted.

The LUBRICATION system delivers oil to all moving parts inside the engine. The oil in the crankcase is either circulated by a pump or splashed through the engine. The lubricating oil seals the rings in the cylinder, removes some heat from the piston, crankshaft, and valve train, cushions the shock experienced by the bearings, cleans particles and dirt from the bearings and the cylinder, *and* reduces friction. The importance of keeping the engine properly filled with clean oil is apparent.

The COOLING system removes the excess heat from the engine. The flywheel causes air to circulate over the fins of the engine where the excess heat is removed. This system requires periodic service because accumulated dust and grass can cause the engine to overheat.

The EXHAUST system removes the burned gases from the engine and muffles the noise of combustion. On most units the exhaust system consists of a muffler attached to the engine block. A clogged or internally collapsed muffler can cause a loss of power and poor engine performance.

PRINCIPLES OF COMBUSTION

Figure 3

COMBUSTION

POTENTIAL ENERGY

KINETIC ENERGY

HEAT EXPANDS AIR IN BALLOON

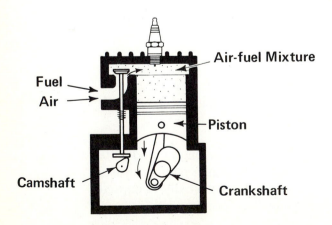

Air-fuel Mixture

Fuel

Air

Piston

Camshaft

Crankshaft

INTAKE OF GAS

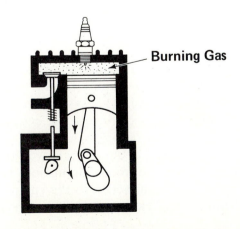

Burning Gas

EXPANSION OF BURNING GAS
PUSHES PISTON

PRINCIPLES OF COMBUSTION

Combustion is defined most simply as *burning*. Combustion in the internal-combustion engine refers to burning that takes place inside the engine.

The fuel stored in the fuel tank represents potential energy. This means that while in the tank little or no energy is released. If a drop of the fuel is placed in a pan and ignited, kinetic energy, or energy that has been released, is formed. The burning fuel gives off heat and light.

Pure fuel (gasoline) will provide heat when burned. If, however, the fuel is mixed properly with air to form a gas, it will give off more heat or burn more efficiently.

If heat is applied to air in a confinement, the air expands. If a balloon is placed over a heat duct, the balloon becomes "larger" because of the "expansion" of the air within. Burning the mixture in the cylinder in like manner expands the gases in the cylinder.

In the internal-combustion engine, the fuel is mixed with the air by the carburetor. Approximately four gallons of air are mixed with each tablespoon of gasoline. The mixture enters the engine as a mist and changes to a gas or vapor when it reaches the hot port area of the engine. The mixture enters the engine because the piston creates a partial vacuum (suction) as it moves downward on the intake stroke.

At the end of the intake stroke, the intake valve closes, thus trapping the gases in the cylinder. As the piston moves toward TDC (top dead center), the trapped gas is compressed.

The compressed gas will burn more violently than if it were not compressed. The violent burning creates heat, which causes expansion of the gas in the cylinder. The expanding gas exerts much pressure, which causes the piston to move down in the cylinder.

The gas acting upon the piston is the point at which the chemical energy (burning gas) is converted into mechanical energy. It is the mechanical energy that is used to turn the crankshaft.

Combustion, then, in terms of a piston engine means that an air-fuel mixture is drawn into the cylinder, is changed to a gas by heat, is compressed by the piston, and is ignited to provide heat that expands the gases in the cylinder. Expansion of the gases in the cylinder pushes the piston, which transmits the force to the connecting rod and crankshaft.

It is important to note that the air-fuel mixture *does not* explode in the engine. Instead, it burns rapidly. If the engine is timed improperly or if the air-fuel mixture is incorrect, the mixture may "explode" or burn too fast. Operating the engine in this condition will produce "knocking" and excess heat that can damage the piston.

ENGINE IDENTIFICATION

Figure 4

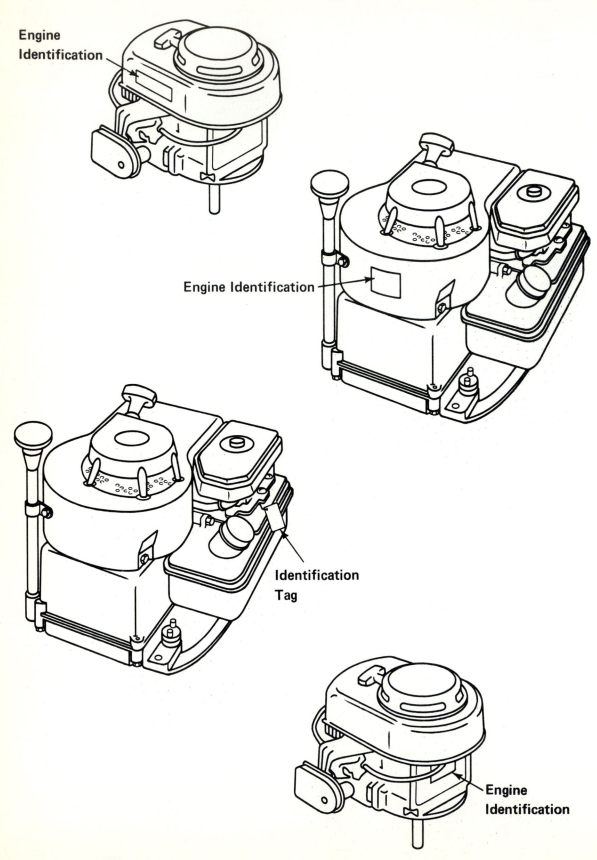

Engine
Identification

Engine Identification

Identification
Tag

Engine
Identification

ENGINE IDENTIFICATION

The identification of the engine is necessary for locating specifications and for purchasing replacement parts. It is always good practice to take the old part along when one goes to purchase a new part.

The engine is usually identified by manufacturer's identification on the engine. This identification usually includes the model information and in some cases the serial number and type.

The identification plate or information may be located on the engine block or on the flywheel shroud. In some cases the number will be stamped on the shroud and no identification tag will be used.

Copy this identification information for your engine in the space below. Use it for future reference when referring to specifications or ordering parts.

Manufacturer _____   Model No. _____

Serial No. _____   Type _____

FOUR-STROKE CYCLE THEORY AND INTAKE STROKE

Figure 5

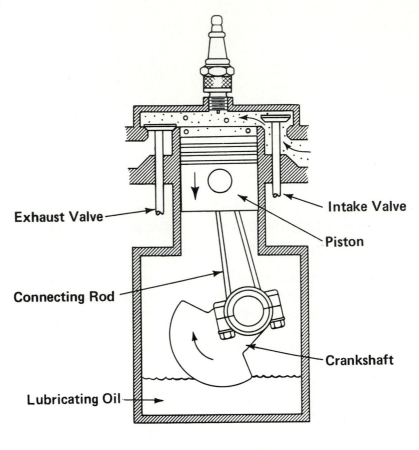

Exhaust Valve

Intake Valve

Piston

Connecting Rod

Crankshaft

Lubricating Oil

INTAKE STROKE

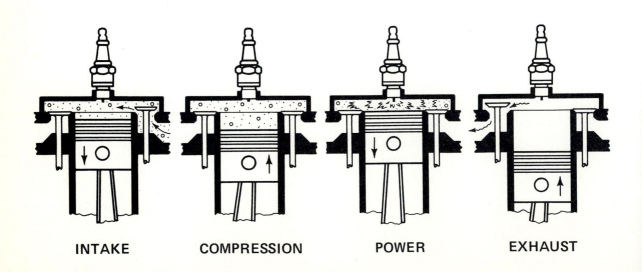

INTAKE   **COMPRESSION**   **POWER**   **EXHAUST**

26

FOUR-STROKE CYCLE THEORY
AND INTAKE STROKE

"Four-stroke cycle" means that it requires four strokes of the piston to complete one cycle.

Each time the piston moves from the top of the cylinder to the bottom of the cylinder it completes one stroke. Similarly, each time the piston moves from the bottom of the cylinder to the top of the cylinder it completes one stroke.

The term "cycle" means completion of four movements or strokes of the piston before it repeats a stroke.

To complete one stroke either from bottom dead center (BDC) to top dead center (TDC) or from TDC to BDC involves moving the crankshaft one-half revolution, or 180°. Because four strokes are required to complete one cycle, it can be seen that the crankshaft completes two revolutions or 720° of rotation during one cycle.

In one cycle the intake valve opens one time and the exhaust valve opens one time. This means that the camshaft, which causes the valves to open, completes but one revolution per cycle.

A reservoir of oil is maintained in the base or crankcase of the engine and provides lubrication for the internal components.

INTAKE STROKE
During the intake stroke the piston is moved from TDC to BDC. This causes a partial vacuum or suction inside the cylinder. Atmospheric pressure (outside air) rushes toward the partial vacuum in the cylinder. It moves through the air cleaner where dirt is filtered out. From the air cleaner it rushes through the carburetor where it "picks up" some gasoline. This air-fuel mixture then enters the cylinder through the open intake valve.

Near the end of the intake stroke the intake valve closes. When the valve is tightly closed, the air-fuel mixture is trapped in the cylinder. At this point the intake stroke is completed and the piston is at BDC (bottom dead center).

COMPRESSION STROKE

Figure 6

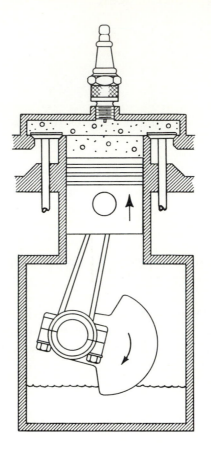

COMPRESSION STROKE

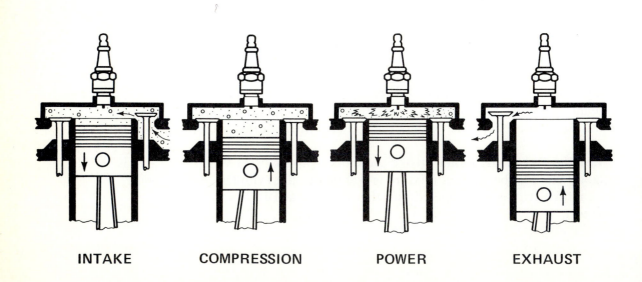

INTAKE   COMPRESSION   POWER   EXHAUST

COMPRESSION STROKE

During the intake stroke an air-fuel mixture was "drawn" into the cylinder by the piston moving from TDC to BDC.

The second stroke in the cycle is the *compression* stroke. The piston is moved from BDC to TDC by the rotating crankshaft. Note that both valves remain closed during this stroke. This means that there is no way that the air-fuel mixture can escape—it is trapped in the cylinder. As the piston moves toward TDC, the air-fuel mixture is compressed. This compression takes place because, as the piston moves toward TDC, the volume of the cylinder decreases.

By compressing the air-fuel mixture, more pressure is created when it is burned.

POWER STROKE

Figure 7

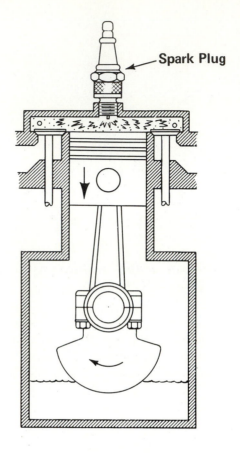

Spark Plug

POWER STROKE

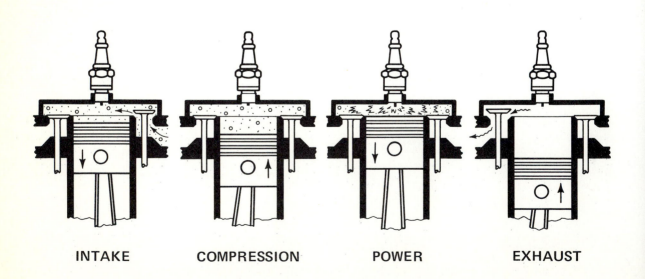

INTAKE   COMPRESSION   POWER   EXHAUST

POWER STROKE

The air-fuel mixture was compressed as the piston moved from BDC to TDC during the compression stroke.

Just before the piston reached TDC on the compression stroke, the ignition system created enough voltage to jump the spark plug gap. When the spark jumped the gap of the spark plug, the *power* stroke began.

The spark ignites the air-fuel mixture. As the mixture burns, the gases expand. The expansion of the gases increases the pressure inside the cylinder. Because both valves are closed, no means exist whereby the pressure can leak out. The increasing pressure is exerted on all parts of the inside of the cylinder. The piston is the only part that can move. The pressure forces the piston to move toward BDC. As the piston moves, the force is transmitted to the crankshaft by the connecting rod.

The power stroke is the only stroke in the cycle that produces usable energy. The power stroke causes the crankshaft to revolve, which makes the engine "run."

EXHAUST STROKE

Figure 8

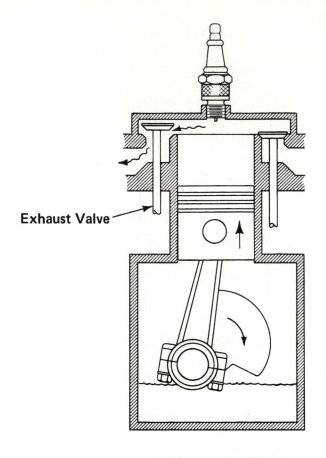

Exhaust Valve

EXHAUST STROKE

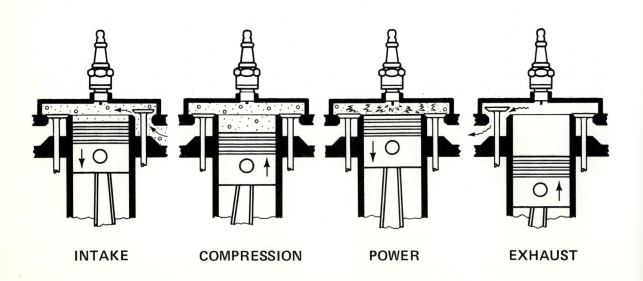

INTAKE   **COMPRESSION**   **POWER**   **EXHAUST**

EXHAUST STROKE

The power stroke moved the piston from TDC to BDC. Near BDC the expansion of gases stopped, which ended the power stroke.

Near BDC of the power stroke the camshaft and valve train cause the exhaust valve to open. As the piston moves from BDC toward TDC, the burned gases are forced out through the open exhaust valve and muffler. This is the *exhaust* stroke. Near the end of the exhaust stroke, the exhaust valve closes and the intake valve opens. This action signals the beginning of another cycle.

FOUR STROKE CYCLE SUMMARY

The sequence of strokes—intake, compression, power, and exhaust— continues to repeat as long as the engine is running.

The important points of the four-stroke cycle operation are as follows:

1. Each time the piston moves from one end of the cylinder to the other end a stroke is completed.
2. Four strokes—intake, compression, power, and exhaust—are required to complete one cycle of a piston.
3. To complete one cycle, the crankshaft must make *two* complete revolutions and the camshaft must make *one* revolution.
4. The engine receives power to the crankshaft *only* on the power stroke. The crankshaft "coasts" through the intake, compression, and exhaust strokes.

The fact that the engine crankshaft must "coast" through three strokes requires a flywheel. Once in motion, the flywheel tends to continue to turn. The heavier the flywheel, the greater its tendency to continue to turn after the power causing it to turn is eliminated.

The flywheel on the small single-cylinder engine is relatively heavy because there is *only* one piston delivering power to the crankshaft. On many rotary-type lawnmowers the blade of the mower also helps the flywheel to maintain rotation of the crankshaft. Many of these types of engines run roughly if the mower blade is not bolted to the crankshaft.

TWO-STROKE CYCLE THEORY

Figure 9

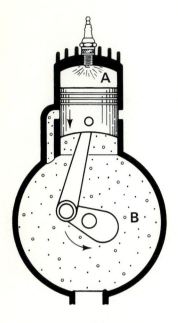

I. DOWN STROKE
(POWER)

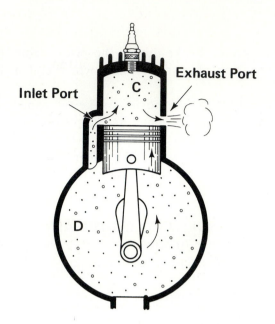

II. END OF DOWN STROKE
BEGINNING OF UP STROKE

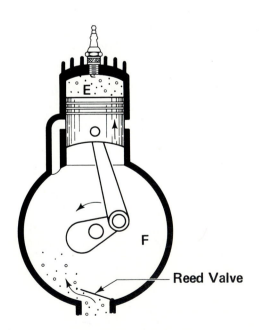

III. UP STROKE (COMPRESSION)

TWO-STROKE CYCLE THEORY

Two-stroke cycle means that two strokes of the piston complete one cycle. In other words, the piston must move from TDC to BDC and then back to TDC in order to complete a cycle. The two-stroke cycle engine completes a cycle each time the crankshaft completes one complete revolution.

Part I in Figure 9 shows the piston moving down from TDC. The air-fuel mixture has been compressed (or squeezed) above the piston, and the ignition system has created a spark at the spark plug. The spark has jumped the gap at the spark plug igniting the air-fuel mixture.

As the gases burn (A), they expand, thereby pushing the piston. As the piston moves down, the air-fuel mixture (B) (below the piston) in the crankcase is compressed.

At the end of the down stroke (Figure 9, Part II) the compressed gases in the crankcase (D) flow up to the cylinder through the intake port. The exhaust gases are released from the cylinder at the end of the power stroke when the piston has passed the exhaust port (uncovering the port), which allows them to escape. The incoming intake gases help clean the remaining exhaust gases from the cylinder (C). Note that at BDC both ports are open.

When the piston begins the up stroke, it covers both the intake and exhaust ports, thereby sealing the cylinder and trapping the intake gases. As the piston continues to move upward, the intake gases are compressed.

Part III of Figure 9 shows that while the gases on top of the piston are being compressed (E), a partial vacuum is being created below the piston (F). The crankcase of the two-cycle engine is small so that this partial vacuum can be created. Air rushes in through the carburetor, picks up fuel, and enters the crankcase through the reed valve.

The reed valve is a one-way valve—it allows air-fuel mixture to enter the crankcase but will not let them out once they have entered.

When the piston reaches TDC, ignition takes place and the cycle begins again.

The two-stroke cycle piston engine has activity on both the top and the bottom of the piston. Gases are compressed on both sides; however, those compressed in the crankcase do not burn. Instead, they move through the inlet port to the top side of the piston.

Most two-stroke cycle engines have no reservoir of oil in the crankcase. Lubricating oil is mixed with the fuel. When the air-fuel mixture enters the crankcase, the fuel and air form a vapor while the oil remains in droplets that cling to the components inside the crankcase. These droplets of oil maintain an oil film on the inside of the crankcase and all the moving parts. This film provides the necessary lubrication for the engine.

Questions for Section II

T F 1. For proper combustion, fuel must explode in the cylinder. Page 23.

T F 2. Large drops of fuel are required in the cylinder to produce power. Page 23.

T F 3. The crankshaft receives power during only one of the four strokes of the four-stroke cycle engine. Page 33.

T F 4. By compressing the air-fuel mixture, more pressure is created when it is burned. Page 29.

T F 5. During the power stroke, one valve is open. Page 31.

T F 6. Lawnmower engines may tend to run roughly when operated without a blade. Page 33.

T F 7. A flywheel is needed only to help cool the engine. Page 33.

T F 8. In a two-stroke cycle engine, the air-fuel mixture is compressed first in the crankcase. Page 35.

T F 9. On a two-stroke cycle engine, the air-fuel mixture is admitted to the crankcase through reed valves. Page 35.

10. List the five basic systems of the small gas engine. Page 21.

11. Describe how excess heat is removed from the small gas engine. Page 21.

12. What is the job of the ignition system? Page 21.

13. List the four things you need to know about the engine when referring to specifications or ordering parts. Page 25.

14. List the four strokes of the four-stroke cycle in their correct order. Page 33.

15. What is a stroke? Page 27.

16. What is meant by TDC and BDC? Page 27.

17. How many times does the crankshaft turn to complete the four-stroke cycle? Page 27.

18. Explain how the two-stroke cycle engine is lubricated. Page 35.

Section III
BASIC ENGINE TESTING

Figure 10

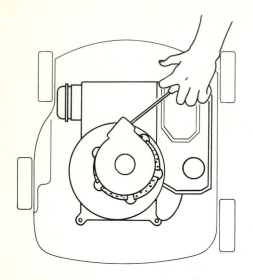

CHECKING COMPRESSION BY
NOTING RESISTANCE WHEN PULLING
STARTER ROPE

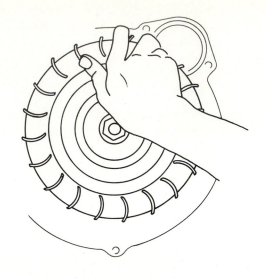

CHECKING COMPRESSION BY
ROTATING FLYWHEEL IN DIRECTION
OPPOSITE NORMAL ROTATION

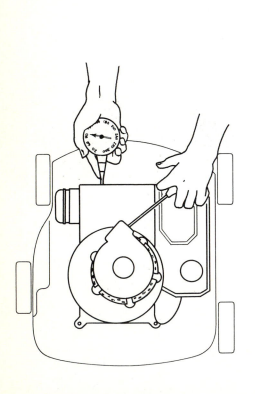

CHECKING COMPRESSION WITH
A COMPRESSION GAUGE

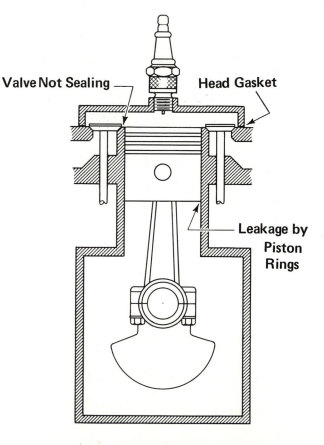

Valve Not Sealing

Head Gasket

Leakage by
Piston
Rings

PLACES WHERE COMPRESSION
AND POWER CAN LEAK

COMPRESSION TESTING

For an internal-combustion engine to run, three conditions must be met. The engine must have the following:

Compression

Air-fuel mixture

Ignition

These three "musts" have to be working together or timed if satisfactory engine operation is to be achieved.

Test 1. "Ground" the spark plug wire by connecting it to the engine. Check the compression by pulling on the starter rope. As the rope is being pulled, one should be able to feel the resistance as the piston comes upon compression. NOTE: Some engines have a built-in compression release that will release compression until the engine attains a proper speed. Usually the speed at which the release disengages occurs at the "end of the pull" of the rope when pulled at a normal starting rate. On engines with a compression release or one that fails the test, test the compression as outlined in Test II or III.

Test II. If compression cannot be felt in Test I, make certain that the spark plug wire is grounded and spin the flywheel by hand in a direction opposite its normal rotation. This can be done by removing the air shroud. When rotated in the opposite direction, the flywheel should rebound as the piston comes upon compression. If the rebound is quick, compression is good.

Test III. A third method of testing compression involves using a compression gauge. The spark plug is removed and the gauge is held tightly in the spark plug hole. The engine is cranked normally and the gauge reading is observed. The normal pressure reading varies with the size of the engine. For the lowest compression engines, the reading should not be below 25 psi (pounds per square inch). Most three-horsepower engines will give a 60–75 psi reading. Refer to the manufacturer's manual for the normal compression pressure of a given engine. NOTE: This test should not be used on engines that are identified as having "easy spin" starting.

Low compression can result from several different faults. If there is a hissing sound while the compression is being checked there may be a leaking head gasket.

Worn or burned valves or valve seats will cause low compression readings.

Worn or improperly seated piston rings will cause compression loss into the crankcase. In extreme cases the engine will "belch" oil and smoke from the crankcase breather while it is running.

Low compression will usually necessitate engine disassembly. In some cases a partial disassembly (head gasket and burned valves) will be all that is necessary. In other cases it will be necessary to overhaul the engine. Major engine service is covered in the Overhaul section of this book.

SPARK TESTING

Figure 11

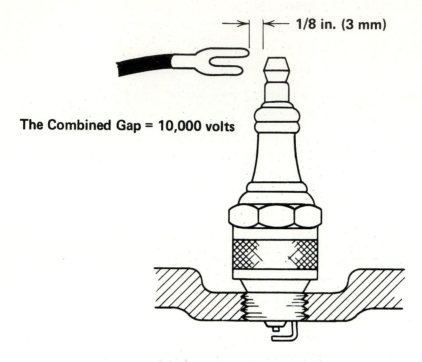

1/8 in. (3 mm)

The Combined Gap = 10,000 volts

QUICK CHECK METHOD

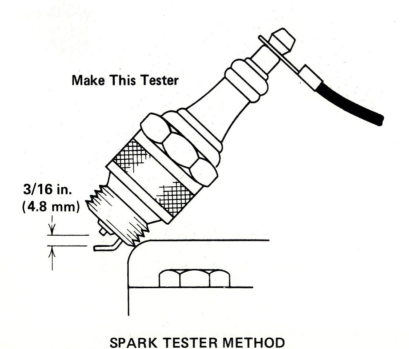

Make This Tester

3/16 in. (4.8 mm)

SPARK TESTER METHOD

SPARK TESTING

To fire the spark plug consistently, approximately 6,000 volts is needed at idle and up to 10,000 volts at high speed or when under load. All of the ignition components must be in good condition in order to deliver this high-voltage spark.

The flywheel magneto ignition consists of the spark plug, high-tension wire, coil, breaker points, and condenser, or solid state components. A magnet attached to the flywheel supplies the energy to the coil. Failure of any of these components can affect the spark. Some of the components may be checked separately, for example, the coil, spark plug, and the flywheel magnet.

If all the components are functioning properly, the magneto should be able to produce a three-sixteenth-inch spark when cranked rapidly. Either of the following methods may be used to check the spark; the first method is a little quicker to perform and the second method is a little more precise.

QUICK CHECK METHOD

1. Carefully remove the spark plug wire from the spark plug. Pull the boot (rubber cover) back so that the metal connector is exposed.
2. Grasp the spark plug wire by the insulation and hold the spark plug wire connector one-eighth inch from the spark plug terminal. Fold a hand towel and place it between your hand and the engine head so that your hand can rest firmly against the engine to hold the wire steady. The cloth towel will help prevent accidental shock.
3. Crank the engine vigorously. *CAUTION:* THE ENGINE COULD START—BE PREPARED! The spark should jump the combined gap easily *and steadily*. It should jump the one-eighth-inch (3 mm) gap you are creating outside the engine and the gap of the spark plug. Failure to pass this test could be the fault of the spark plug. If the spark will not jump the one-eighth-inch gap or is unsteady, perform the next test.

SPARK TESTER METHOD

A number of spark testers are available, but a very satisfactory one can be made by using a new spark plug. Be sure to select a spark plug that is not the resistor or booster gap type. Adjust the tester electrodes to make a three-sixteenth-inch (4.8 mm) gap.

Connect the spark plug wire to the tester and hold the tester firmly against the engine. Make sure that the tester is making contact with the bare metal of the engine because paint may insulate the tester from the engine.

Spin the engine rapidly. A steady spark across the gap indicates that the ignition system is functioning satisfactorily. In this method the spark is jumping only one gap and the results are more reliable. Also, the engine will spin much easier and there is no chance of the engine's starting with the plug removed. If the plug is removed, see the section on Spark Plug Service.

BASIC CARBURETOR ADJUSTMENTS

Figure 12

NOTE: OBSERVE CAUTION AROUND DRIVE MECHANISMS, BLADES, AND BELTS WHEN MAKING CARBURETOR ADJUSTMENTS!

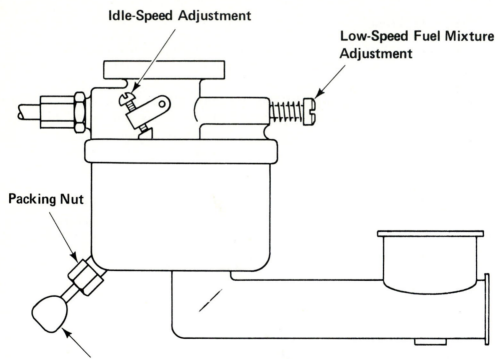

Idle-Speed Adjustment

Low-Speed Fuel Mixture Adjustment

Packing Nut

Load-Speed or High-Speed Mixture Adjustment Needle Valve

Note: Some manufacturers identify this as the "main" jet.

BASIC CARBURETOR ADJUSTMENTS

There are up to three basic adjustments necessary on most carburetors. These adjustments are made in the order listed below. The needle valve seats are made of a soft material like brass or plastic and can be easily damaged. When turning the needle valve in, take care not to damage the soft brass seat. Loosening the packing nut will allow the screw to turn easier. If the fuel leaks around the adjusting screw, tighten the packing nut (see carburetor illustration).

THE LOAD-SPEED AIR-FUEL MIXTURE

Start the engine and allow it to run at least three minutes to warm up before beginning carburetor adjustment. If the engine will not start, carefully turn the load-speed needle valve *in* (clockwise) until it closes. *Do not force it!* Back the screw out approximately one and one-half turns. This is not a correct adjustment, but it is rich enough to start most engines.

If the engine surges or runs roughly and emits black smoke from the exhaust pipe during warm up, the mixture is too rich. Turn the needle valve *in* (clockwise) slowly until the engine smooths out. If the engine surges and runs roughly without emitting black smoke, turn the mixture screw *out* (counterclockwise) to smooth out engine performance. Ignore blue smoke coming from the engine exhaust pipe at this time. It is the result of oil burning in the engine and has no effect on the carburetor adjustment unless it is excessive.

After the engine is warmed up, bring the engine to normal load-speed— near wide open throttle. *Caution:* Do not operate the engine at full throttle when it is not under normal load because the excessive speed will damage the engine. Turn the needle valve *in* until the engine begins to surge or run unevenly. This is now a lean adjustment. Turn the needle valve *out* until maximum smoothness and top speed are reached. Turn the needle valve slowly back and forth to find top speed without moving the throttle. If the engine will be pulling a heavy load, for example, a tiller, turn the needle counterclockwise up to one-eighth turn to slightly enrich the mixture for maximum power.

THE IDLE-SPEED AIR-FUEL MIXTURE

If the engine has an idle-fuel mixture adjustment, it should be adjusted after the load-speed mixture has been set and with the engine at near correct idle speed (fast idle). Again, turn this mixture screw *in* (clockwise) until the engine begins to run roughly and then back out to maximum speed. Move the adjustment screw back and forth slowly until maximum speed and smoothest operation are obtained.

THE IDLE-SPEED ADJUSTMENT

The idle-speed adjustment is a stop screw on the throttle shaft that prevents the throttle from closing farther. Single-cylinder, air-cooled engines idle much faster than multicylinder engines. Normal idle speed for engines of less than ten horsepower is 1,300 to 1,800. If the engine will not accelerate properly from idle after the idle speed is set as described above, the idle speed may be too slow. An engine that idles too slowly will overheat because the air flow through the cooling fins will not be sufficient to cool the engine.

Questions for Section III

T F 1. The spark plug should be grounded to the engine when testing compression. Page 39.

T F 2. A magneto should be able to produce a spark across a three-quarter inch gap. Page 41.

T F 3. A leaking head gasket can cause low compression. Page 39.

T F 4. A worn connecting rod bearing can cause low compression. Page 39.

T F 5. A spark tester can be made using a new spark plug with a wide gap. Page 41.

T F 6. The order in which carburetor adjustments are made is not important. Page 43.

T F 7. Carburetor adjustments must be made only after the engine is warmed up. Page 43.

T F 8. Blue smoke is a result of an incorrect fuel mixture. Page 43.

T F 9. Auto engines must idle faster than small gasoline engines. Page 43.

T F 10. Needle valve seats are very hard. Page 43.

11. Why is it best to turn the engine backward for compression testing? Page 39.

12. List three possible causes of low compression. Page 39.

13. In this section, we checked the three essential elements for combustion. List the three elements. Page 39.

14. Using a compression gauge, you would expect a three-horsepower engine to give a reading of about _____ psi. Page 39.

15. What is indicated by low compression accompanied by blue smoke escaping from the crankcase breather when the engine is running? Page 39.

16. What is the voltage required to fire the spark plug at idle speed? Page 41.

17. List the parts of the flywheel magneto ignition. Page 41.

18. List the three basic adjustments that can be made on most small gas engine carburetors. Page 43.

19. Most small gas engines will start if the adjustment screws are opened about _____ turns. Page 43.

20. If the engine runs roughly and emits black smoke, the fuel mixture is probably too _____ and the needle valve should be turned _____ . Page 43.

Section IV

LUBRICATION
AND
COOLING SYSTEMS

LUBRICATION

Figure 13

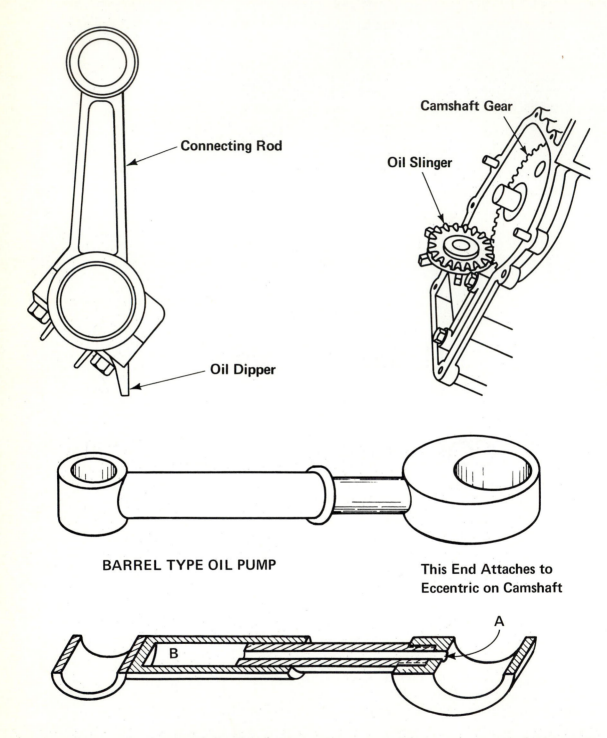

Connecting Rod

Oil Dipper

Camshaft Gear

Oil Slinger

BARREL TYPE OIL PUMP

This End Attaches to Eccentric on Camshaft

A

B

OIL ENTERS AND LEAVES THE PUMP AT POINT A. THE OIL FLOW IS GOVERNED BY OIL PASSAGES IN THE ECCENTRIC. AS THE PUMP EXTENDS, OIL FLOWS INTO CHAMBER B. WHEN THE PUMP PLUNGER IS FORCED IN, THE OIL IS PUSHED OUT OF CHAMBER B, THROUGH PASSAGE A, AND IS DIRECTED THROUGH THE CAMSHAFT.

LUBRICATION

The oil in the crankcase serves five functions:

1. It helps cool the engine by removing heat from the cylinder and moving parts.
2. It cleans by rinsing particles from the cylinder and moving parts.
3. It seals the rings to the cylinder wall.
4. It reduces friction by serving as a slippery film between all moving parts.
5. It protects the machined parts from rust and corrosion.

The oil is circulated in the engine by several methods. Some engines use a dipper attached to the connecting rod, which dips into the oil as the connecting rod moves. The oil is splashed so that all points within the block are constantly sprayed with oil.

The oil slinger used on some engines is driven by the camshaft. As the slinger rotates, "ears" on the slinger throw oil throughout the inside of the engine.

Various styles of oil pumps have been used in the lubrication systems of small engines. The pumps normally are submersed in oil and direct oil to the connecting rod and main bearings by tubes or orifices in the discharge tube. The other parts of the engine are lubricated by the oil that is "thrown off" the connecting rod and crankshaft.

When an engine that employs an oil pump is being overhauled, the pump should be checked carefully for wear. If the pump shows wear, replace the main components or the complete pump assembly.

OIL AND OIL CHANGE

The oil level should be checked before starting the engine and after every 5 or 6 hours of operation. The oil level should be maintained to the full mark as recommended by the engine manufacturer. NOTE: Some engines employ an oil sensor that disables the ignition when the oil level is too low. When checking or adding oil, care should be taken to prevent any dust or dirt from entering the oil filler. The engine oil should be changed when it is warm so that all the oil will drain from the engine. Change the oil at least after every 25 hours of engine operation.

Use the weight and grade oil recommended by the manufacturer. The use of additives is not necessary unless specified by the manufacturer.

If manufacturer data are not available, use the following guide in selecting oil:

Service rating	SC, SD, SE, or SF (SF is the highest rating)
Type	Detergent
Summer	SAE 30 or 10W-30
Winter	SAE 10W or 10W-30

COOLING SYSTEM

Figure 14

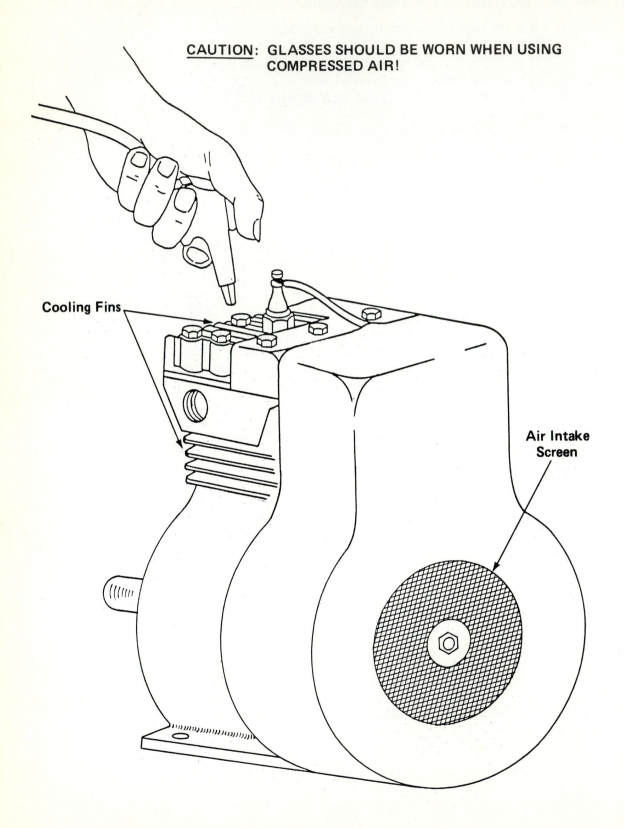

CAUTION: GLASSES SHOULD BE WORN WHEN USING COMPRESSED AIR!

Cooling Fins

Air Intake Screen

COOLING SYSTEM

Excess heat from the small engine is removed by air circulated over the engine. Air is drawn in through the screen in the shroud that covers the flywheel. The spinning flywheel throws the air outward and the shroud directs it across the engine.

It is very important that the air intake screen and the fins on the engine be kept clean. If dust and grass are permitted to build up over these areas, the air flow is restricted. Such restriction of the air flow can cause the engine to run "hot."

Blockage of just one passage can create a "hot spot" that can cause scratched piston rings, a scored cylinder wall, and scoring of the piston.

The engine shrouding should always be in place when the engine is operated. Without the shrouding, the air is not directed over the engine and the engine will overheat.

The screen over the air inlet area should be brushed frequently with a stiff-bristled brush to remove accumulated grass and dust.

The shroud should be periodically removed from the engine and the accumulated dust and chaff should be removed by blowing around the fins with compressed air.

When the engine is being serviced, care must be taken to prevent breakage of the cooling fins on the engine block and cylinder head. These fins provide necessary "extra" surface area to remove the excess heat from the engine.

Broken flywheel fins or blades reduce the amount of air circulated over the engine and cause the flywheel to be out of balance. If any flywheel fins are broken or missing, the flywheel should be replaced.

The fuel system and ignition timing can also affect engine cooling. If the fuel mixture is lean, the engine will tend to run hot. Late ignition timing will also increase the operating temperature of the engine.

T F 1. Engine oil should be changed when the engine is still warm. Page 47.

T F 2. Oil additives are necessary to fortify the oil used in the crankcase of small gas engines. Page 47.

T F 3. The light metal shrouding around the engine is for appearance only and should be removed for best performance. Page 49.

T F 4. Fuel adjustments and engine timing can affect engine temperatures. Page 49.

T F 5. If there are broken or missing fins on the flywheel, the flywheel should be replaced. Page 49.

T F 6. Engine oil should be checked every 5 or 6 hours of operation. Page 47.

T F 7. A "hot spot" on the cylinder wall can cause scoring of the cylinder. Page 49.

T F 8. The engine fins and the air-intake screen should be frequently cleaned with a brush, vacuumed, or blown with compressed air. Page 49.

9. List the five functions that oil must perform in the engine. Page 47.

10. List the different methods used to circulate oil through the engine. Page 47.

11. Describe how the small engine is cooled. Page 49.

12. Engine oil should be changed after every _____ hours of operation. Page 47.

13. What conditions can cause engine overheating besides problems with the cooling or lubrication system? Page 49.

14. What oil service rating is the best available? Page 47.

Section V

ELECTRICAL FUNDAMENTALS
AND
IGNITION SYSTEMS

Figure 15

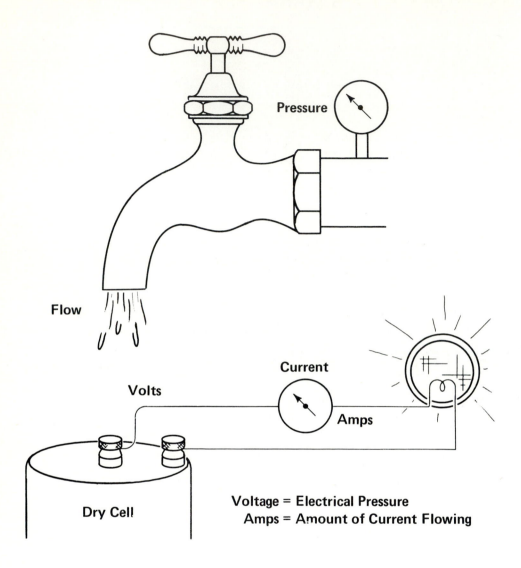

Pressure

Flow

Current

Volts

Amps

Dry Cell

Voltage = Electrical Pressure
Amps = Amount of Current Flowing

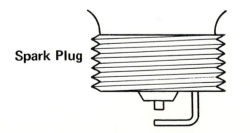

Spark Plug

6,000 Volts Needed to
Fire the Plug at Idle—More
at Higher Speeds and
Greater Loads

ELECTRICAL TERMS

Voltage (volts) is the pressure that causes current to flow. In this sense, it is no different from the pressure that pushes water through a pipe. If you increase the pressure in a water system, more water will flow. If you increase the voltage in an electrical circuit, more current will flow.

Current is the moving of electrons in a wire. This is electricity. Just as water moves in a pipe, electrons flow in a wire. This flow of electrons is called current. The amount of current that is flowing is called *amperes* (amps) and is measured with an ammeter.

If a faucet has 40 pounds of pressure, the amount of water flowing depends on how much the faucet is opened and on the length of the hose that is connected to the faucet. When a 1 1/2-volt battery is used, the amount of current that flows depends on what is connected to the battery. The wire used must be large enough to easily carry that amount of current. Both the headlights and the dash lights on a car operate from 12 volts. Obviously, the headlights allow more current flow and require larger wire. When the voltage is fixed, as when using a battery, the load determines the amount of current that must flow to get the work done. Opening a faucet wider allows more water to flow and does a better job of blasting off dirt. More work is being done. A larger light bulb allows more current to flow and produces more light.

A spark plug simply creates a gap in the electrical circuit. Current flows easily along the wire to the plug but cannot easily jump the gap. A lot of push (voltage) is needed to cause the current to jump. Usually the voltage required to fire the plug at idle speed is about 6,000 volts and may require as high as 10,000 volts on a small engine under load. High-compression auto engines may require in excess of 16,000 volts under load. Current will not flow across the spark plug electrodes until the induced voltage builds high enough to cause the flow to jump the spark plug air gap.

Figure 16

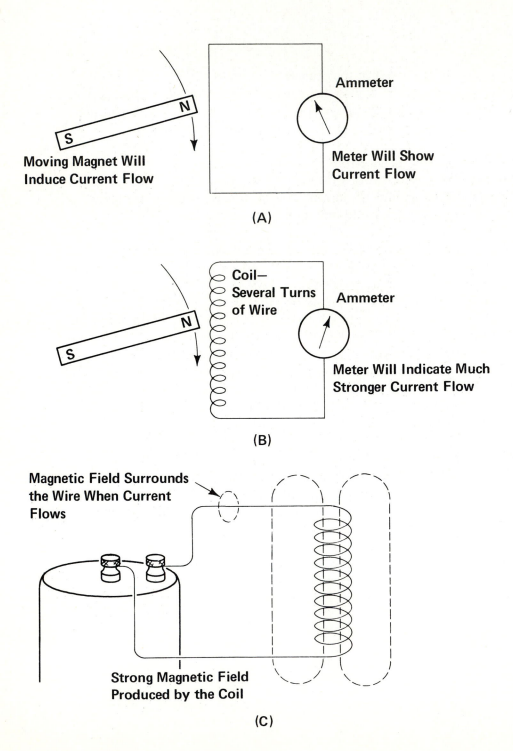

**Moving Magnet Will
Induce Current Flow**

Ammeter

Meter Will Show
Current Flow

(A)

Coil—
Several Turns
of Wire

Ammeter

Meter Will Indicate Much
Stronger Current Flow

(B)

Magnetic Field Surrounds
the Wire When Current
Flows

Strong Magnetic Field
Produced by the Coil

(C)

ELECTRICAL FUNDAMENTALS

The following are basic rules that one must comprehend if the basic magneto is to be understood. Study them carefully.

1. Moving a magnet past a wire causes an electrical current to flow in the wire if it is part of a complete circuit.
2. Moving a magnet past a coil made up of several turns of wire will produce a much stronger current flow because the current induced in each turn of wire will add to the total output.
3. Current flowing in a wire causes a magnetic field to surround the wire. The strength of this field depends directly on the amount of current flowing through it.
4. If the wire is wrapped to make a coil, the strength of the magnetic field of each turn of wire adds to the field next to it, producing a stronger magnet. Adding more turns of wire adds to the strength of the magnetic field.

Figure 16A shows a current meter (ammeter) connected to a loop of wire. Moving the magnet rapidly causes a very small current to flow in the wire, which causes the needle to deflect. Wrapping several turns of wire on a coil makes the current flow much stronger. The current flow induced in each turn of the coil adds to the current flow induced in all of the other turns on the coil to produce a strong current flow.

If a battery is connected to a complete circuit, the current flowing will cause a magnetic field to surround the wire. The more current that is flowing, the stronger the magnetic field. By passing the current through a coil made up of several turns of wire, the magnetic strength of each loop of wire adds to the others making a stronger magnetic field (Figure 16C).

SPARK PLUG SERVICE

Figure 17

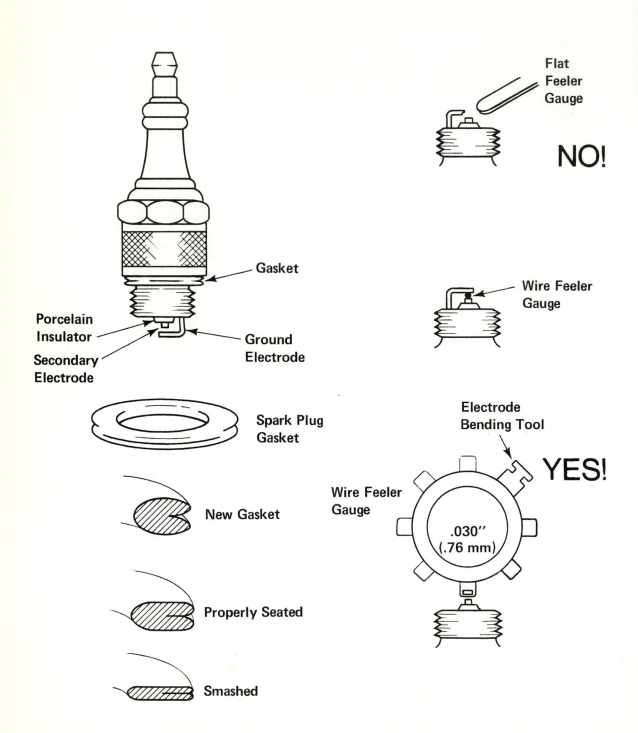

Flat Feeler Gauge

NO!

Gasket

Porcelain Insulator

Secondary Electrode

Ground Electrode

Wire Feeler Gauge

Spark Plug Gasket

New Gasket

Properly Seated

Smashed

Electrode Bending Tool

Wire Feeler Gauge

.030'' (.76 mm)

YES!

SPARK PLUG SERVICE

First remove the spark plug wire and attach it to a good engine ground. Use a deep-well spark plug socket or box end wrench to loosen the plug. Use care not to damage the plug. Loosen the plug about one turn. Spin the engine four to five revolutions to blow out the carbon that may have been knocked loose when the plug was loosened. Blow away any trash, dirt, and caked oil from the area around the plug on the outside of the engine. Remove the spark plug from the engine.

INSPECT THE PLUG
The condition of the plug gives some clues to the condition and operation of the engine.

Normal. If the plug has only slightly worn electrodes and a very light coating of tan or gray deposits, it may be cleaned, regapped, and reinstalled.

Carbon Fouled. Fluffy, black deposits are a result of over-rich carburetion or excessive idling. If electrode wear is only slight, the plug may be cleaned and reinstalled. Be sure to correct the condition that caused the carbon fouling. BE SURE THAT IT IS THE CORRECT PLUG FOR THE ENGINE.

Oil Fouled. Wet, black, oily deposits indicate that oil is leaking past the rings or valve stems. Unless the engine condition allowing oil to enter the combustion chamber is corrected, a new plug will soon become fouled too. An engine overhaul may be necessary to obtain satisfactory service.

Blistered Insulator. A burnt or blistered insulator is the result of overheating. Using the wrong spark plug, low-octane fuel, incorrect timing, bad valves, and cooling system obstruction are common causes of overheating. Correct the cause of overheating and replace with a new spark plug.

DO NOT CLEAN SPARK PLUGS WITH ABRASIVE CLEANERS
Blasting the plug with abrasives quickly erodes the insulator nose and may change the heat range of the plug. Any deposits that are not readily removed with a wire brush may be scraped away lightly with a pen knife. If excessive deposits are present or if any question exists as to whether or not the plug should be replaced, replace it. A spark plug is a minor expense for the major role it plays in engine performance.

GAP THE NEW SPARK PLUG
New spark plugs are not preset. Check the engine manufacturer's specifications for the correct setting. The correct setting is .025"–.030" (.64–.76 mm) for most small gas engines. Use a wire spark plug feeler gauge because a flat feeler gauge can give erroneous readings, as shown in Figure 17.

REPLACE SPARK PLUG
If the new plug does not screw in easily by hand, clean the threads so that the plug does screw in easily. Then tighten with a torque wrench, if available, to the manufacturer's specifications. If a torque wrench is not available, tighten down finger tight against the new gasket and then turn another one-third turn. This should partially collapse the gasket but not completely smash it. Do not overtighten. Remember that the spark plug has steel threads and the engine head is most likely aluminum or cast iron.

THE IGNITION CIRCUIT

Figure 18

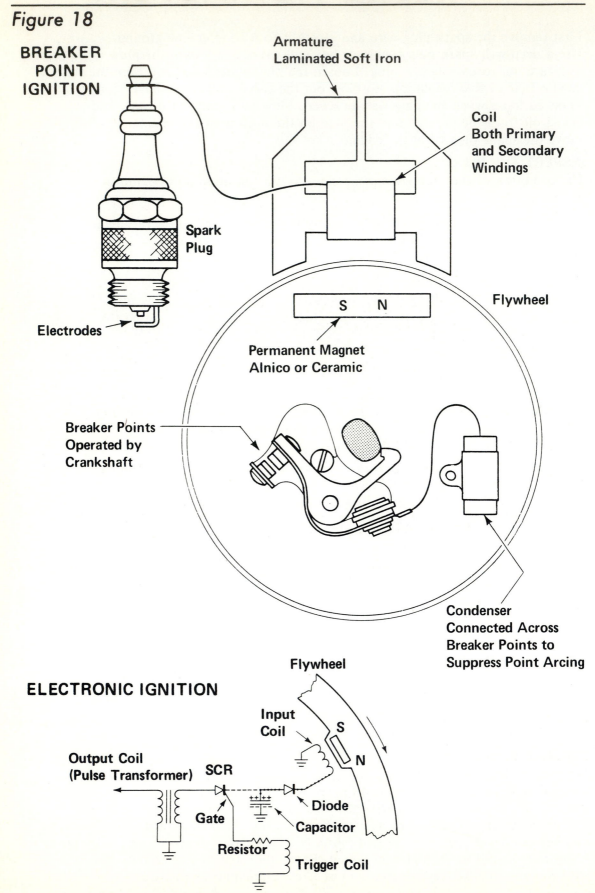

BREAKER POINT IGNITION

Armature
Laminated Soft Iron

Coil
Both Primary
and Secondary
Windings

Spark
Plug

Electrodes

S   N

Flywheel

Permanent Magnet
Alnico or Ceramic

Breaker Points
Operated by
Crankshaft

Condenser
Connected Across
Breaker Points to
Suppress Point Arcing

ELECTRONIC IGNITION

Flywheel

Input
Coil

S

N

Output Coil
(Pulse Transformer)

SCR

Gate

Diode

Capacitor

Resistor

Trigger Coil

THE IGNITION CIRCUIT

The purpose of the ignition circuit is to create a spark to ignite the compressed air-fuel mixture at exactly the right time. In order to do this, a very high voltage is needed to cause an electric current to jump the gap between the spark plug electrodes. The duration or intensity of this small lightning bolt is not important because the compressed air-fuel mixture is highly volatile and will be readily ignited if the conditions are correct.

As the magnet in the flywheel passes the coil armature, the moving magnet sets up a current flow in the primary winding of the coil. This current is also traveling through the breaker points. As the breaker points open, or the electronic ignition unit switches, interrupting the current flow, the electrical power is transferred magnetically to the coil's secondary winding at a very high voltage. This voltage travels to the spark plug where it jumps the spark plug electrodes igniting the air-fuel mixture. On engines with ignition points, the points open, a small spark occurs across the points. Because the points open several thousand times each minute, they would soon burn up. The condenser helps absorb this spark and thus acts to increases point life.

THE BREAKER POINT IGNITION SYSTEM

The FLYWHEEL MAGNET moves rapidly past the coil assembly, causing current flow in the coil primary winding.

The ARMATURE-COIL ASSEMBLY consists of the laminated steel armature and the primary and secondary winding.

The BREAKER POINTS are operated by the crankshaft. They close just before the flywheel magnet passes the coil completing the primary winding circuit. The points open just as current flow in the primary is at maximum, causing the power to be transferred magnetically (induced) in the secondary at a much higher voltage to fire the plug.

The CONDENSER is connected across (in parallel with) the points to help reduce arcing of the points, which increases point life.

THE ELECTRONIC IGNITION SYSTEM

On engines equipped with electronic ignition, a trigger coil activates an electronic switch (SCR), which controls the current flow in the coil primary winding. The SCR operates in a fashion similar to the ignition points.

IGNITION PRIMARY CIRCUIT (BREAKER POINT IGNITION SYSTEM)

Figure 19

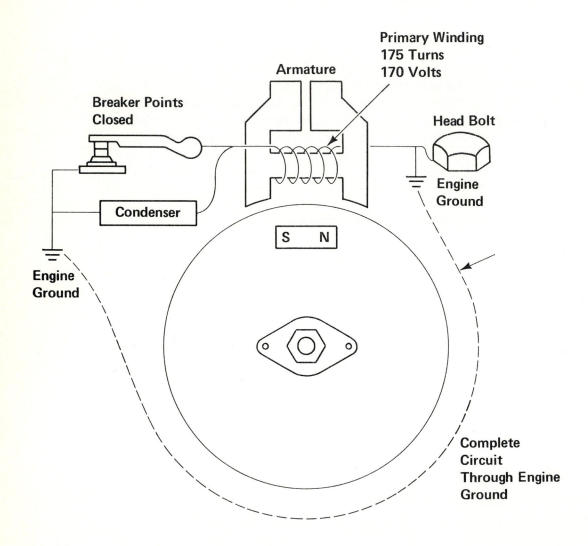

Breaker Points Closed

Armature

Primary Winding
175 Turns
170 Volts

Head Bolt

Engine Ground

Condenser

S N

Engine Ground

Complete Circuit Through Engine Ground

IGNITION PRIMARY CIRCUIT (BREAKER POINT IGNITION SYSTEM

The primary or low-voltage circuit consists of the flywheel magnet, breaker points, condenser, and the coil primary winding. The primary winding usually consists of less than 200 turns of coated copper wire wound on the laminated armature. Note that one lead is connected to the armature on one of the mounting bolts, which connects it to the engine ground. The other lead from the primary winding goes to the breaker points. The condenser is connected across or in parallel with the breaker points. Its only purpose is to prevent arcing of the points, thus increasing point life.

Notice that when the breaker points are closed, there is a complete circuit from the engine ground through the coil primary winding, through the breaker points, to the engine ground. Because both ends of the coil primary are now connected to the engine ground, a complete circuit now exists. Current will flow through the ground as if the two ends of the primary winding were connected directly together.

As the piston is nearing TDC on the compression stroke, the breaker points close, making a complete primary circuit. The flywheel magnet passes rapidly bringing primary current to maximum. Because the current flow in the primary winding is at maximum, a strong, concentrated magnetic field will surround the primary winding. Just before the piston reaches TDC (the manufacturer determines the exact number of degrees before TDC), the breaker points open to stop all current flow in the primary circuit. This causes the magnetic field to suddenly collapse. The condenser absorbs a small amount of the current flow in order to prevent point arcing as the points open.

PRODUCING SECONDARY VOLTAGE

Figure 20

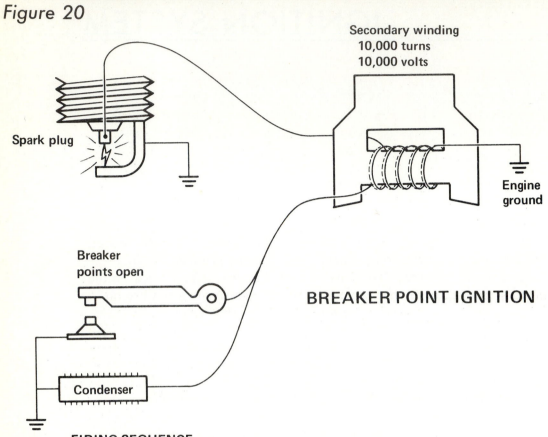

Secondary winding
10,000 turns
10,000 volts

Spark plug

Engine ground

Breaker points open

BREAKER POINT IGNITION

Condenser

FIRING SEQUENCE

1. The points close, making the primary a complete circuit.
2. The flywheel magnet passes, creating current flow in the primary winding.
3. The current flow in the primary winding produces a strong, concentrated magnetic field which surrounds both the primary and secondary windings.
4. The points open, collapsing the magnetic field across the secondary winding.
5. The collapsing magnetic field induces several thousand volts in the secondary winding to fire the spark plug.

ELECTRONIC IGNITION

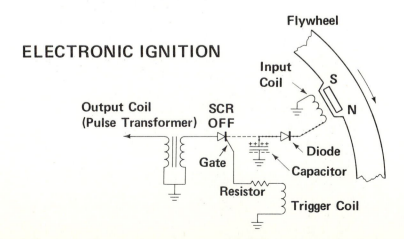

Flywheel

Input Coil

Output Coil
(Pulse Transformer)

SCR OFF

Gate

Diode

Capacitor

Resistor

Trigger Coil

PRODUCING SECONDARY VOLTAGE

BREAKER POINT IGNITION SYSTEM

The coil's secondary winding is wrapped around the primary winding. One end of the secondary winding is connected to the ground, usually along with the lead from the primary winding, which goes to the ground. The other secondary lead goes to the spark plug. The secondary winding consists of approximately 10,000 turns of very fine wire, usually about 60 times more turns than on the primary winding. The wire need not be large because we are concerned in the secondary with creating a very high voltage to push current across the spark plug gap. A very small amount of current is needed.

Because the secondary winding is wound around the primary winding, the magnetic field surrounding the primary winding is also surrounding and saturating the secondary winding. When the points open, the primary magnetic field collapses rapidly. The field collapses across the secondary winding. The collapsing magnetic field is a rapidly moving magnet moving past the secondary winding. This collapsing magnet tries to induce current flow in the secondary winding but because of the spark plug gap, a complete circuit does not exist. Because current cannot flow, voltage builds up until it is great enough to jump the spark plug gap, that is, if the coil can build up enough voltage! This is why the secondary winding is made up of several thousand turns of wire. A small voltage push is created in each turn of wire by the collapsing magnetic field. The voltage of all the turns is added together to create from 10,000 to 12,000 volts to jump the spark plug gap. The spark occurs at the plug when the points open, collapsing the primary magnetic field across the secondary winding, which produces the voltage to push current across the spark plug gap.

ELECTRONIC IGNITION SYSTEM

In the electronic ignition systems, the ignition points and condenser are replaced by electronic components, which serve the same purpose as the ignition points and condenser—they control the flow of current in the primary winding of the pulse transformer. The secondary winding of the pulse transformer is wound around the primary winding in a manner similar to the ignition coil used in the breaker point ignition system. The collapsing magnetic field in the primary winding induces a current flow in the secondary winding just like the induction that takes place in the breaker point ignition system. The operational theory of the electronic ignition is presented in detail on page 79.

FLYWHEEL REMOVAL

Figure 21

CAUTION: SAFETY GLASSES SHOULD BE WORN WHEN USING
IMPACT TO REMOVE A FLYWHEEL

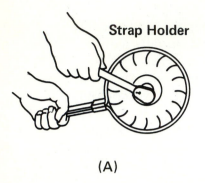

Strap Holder

(A)

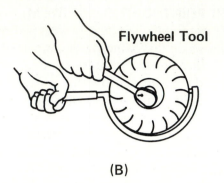

Flywheel Tool

(B)

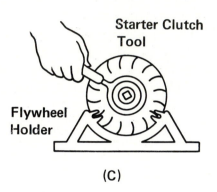

Starter Clutch Tool

Flywheel Holder

(C)

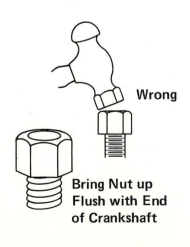

Wrong

Bring Nut up
Flush with End
of Crankshaft

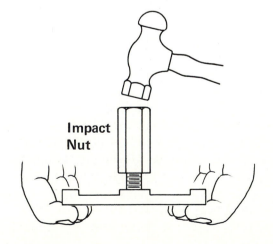

Impact
Nut

(D)

FLYWHEEL REMOVAL

In order to service the points on most models, the flywheel must be removed. Flywheels are usually made of aluminum or cast iron and can be easily warped or cracked. The utmost care must be used in removing the flywheel. Because the flywheel magnet must travel very close to the armature, a warped or cracked flywheel will cause a weak spark or no spark at all. On lawnmowers with brake units which contact the underside of the flywheel, the brake spring tension should be released before removing the flywheel.

The flywheel is fitted on the tapered crankshaft with a key. The flywheel is pressed onto the crankshaft taper and held there by a nut or a threaded starter clutch. The first job is to remove the nut or starter clutch. Use a belt type starter holder as shown in Figure 21A or a flywheel tool (Figure 21B and 21C) to hold the flywheel to remove the nut. DO NOT hold the flywheel with a screwdriver through the cooling fins. If the cooling fins are broken off, serious loss of cooling capacity will occur and the flywheel will become unbalanced and cause vibration and bearing damage. DO NOT try to hold the crankshaft on the power take-off end with a pipe wrench, vise grip, or similar tool because damage to the crankshaft could occur.

Use a socket to remove the nut as shown in Figure 21A. If necessary to remove the starter clutch, use a special starter clutch tool. DO NOT use a hammer on the ears of the starter clutch. If the special tool is not available, make a tool that will engage at least two of the ears on the starter clutch. On some engines, the flywheel nut has left-hand threads. Check threads carefully if there is any doubt about which way to turn the nut. After the nut is removed, the next job is to free the flywheel from the crankshaft taper. Pulling or prying will result in a damaged flywheel. The shock of a sharp blow from a hammer is the best way to loosen the taper fit.

NEVER STRIKE THE CRANKSHAFT WITH A HAMMER
Hitting the crankshaft directly with a hammer will result in damaged threads, a broken crankshaft, or a swollen starter clutch shaft. Also, when shock is used to remove the flywheel, care should be taken that the impact is not placed on the engine's main bearings. The best method is for one person to hold the engine approximately one-fourth inch above the workbench. The engine should be held by gripping the flywheel firmly as shown in Figure 21D. Strike the impact nut with a hammer to deliver the shock to the crankshaft end. Strike the impact nut with a strong, solid blow. One good blow will do what several light taps cannot do. HIT IT!

WEAR SAFETY GLASSES OR FACE SHIELD
Fragments from either the hammer or the impact nut are extremely dangerous.

INSPECT FLYWHEEL AND KEY

Figure 22

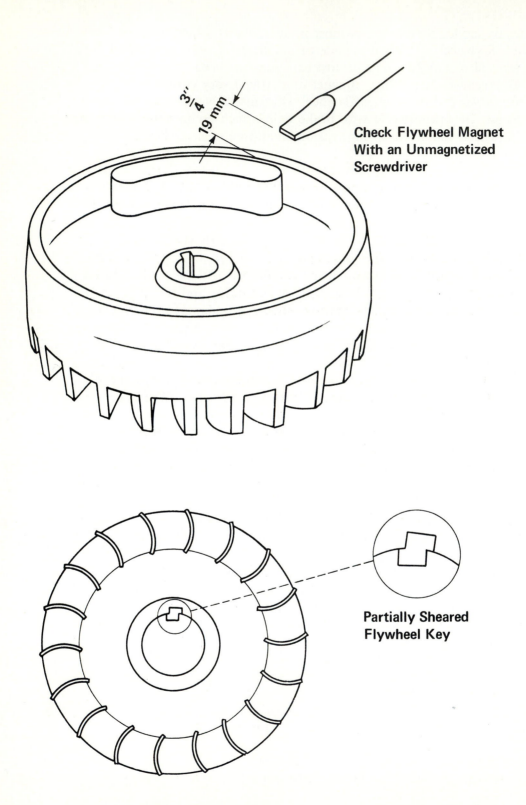

3/4"
19 mm

Check Flywheel Magnet
With an Unmagnetized
Screwdriver

Partially Sheared
Flywheel Key

INSPECT FLYWHEEL AND KEY

Handle the flywheel carefully! Do not drop it or lay it near large motors or electromagnets such as a generator armature growler. Be sure to locate the key that keyed the flywheel to the crankshaft and store it where it will not be lost.

CHECK THE FLYWHEEL FOR CRACKS

Check the surface of the flywheel for cracks. Check for cracks inside the hub near the keyway. If the flywheel is cracked or if the keyway is worn excessively, the flywheel should be replaced.

CHECK THE FLYWHEEL MAGNET

Because the flywheel magnet is the source of power for the flywheel magneto ignition system, it must have full strength. Hold an unmagnetized screwdriver 3/4" (19 mm) from the flywheel magnet. The screwdriver should be strongly attracted to the magnet. If possible, compare with a new flywheel or other good flywheel magnet. Some flywheel magnets are replaceable, but on others the whole flywheel must be replaced. Newer flywheel magnets seldom lose their magnetism unless dropped, heated, or exposed to strong AC magnetic fields.

CHECK THE FLYWHEEL KEY

Some engines, especially lawnmower engines, use an aluminum key to keep the flywheel properly secured to the crankshaft. This is to help absorb the flywheel momentum in case the blade strikes an unmovable object. With such a sudden stop, the momentum of the flywheel would tend to snap the crankshaft. Sometimes the key will shear or partially shear as shown in Figure 22. Replace the key with a new aluminum key. DO NOT replace an aluminum key with a steel key.

Figure 23

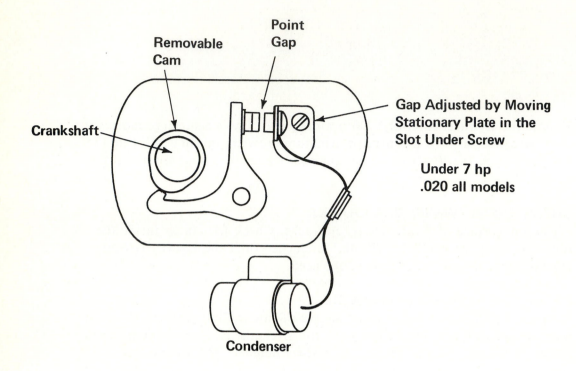

Removable Cam

Point Gap

Crankshaft

Gap Adjusted by Moving Stationary Plate in the Slot Under Screw

Under 7 hp
.020 all models

Condenser

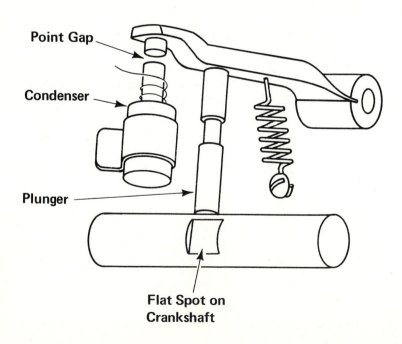

Point Gap

Condenser

Plunger

Flat Spot on Crankshaft

BREAKER POINTS

Breaker points are relatively inexpensive and easy to change. However, considerable care must be used in this task in order to do the job properly. Loss of spark, weak spark, and incorrect spark timing can result from improper point installation. Remember that the spark occurs at the plug when the points open the primary circuit. Setting the breaker points incorrectly will result in improperly set ignition timing.

First, identify the type of breaker point and condenser set used. It will probably be one of the two types shown at the left. Note that on one type the condenser is mounted outside the breaker point box. On the other type the condenser acts as the stationary contact point.

When ordering the new contact points, be sure to have the engine brand, model number, type number, and serial number.

Sometimes it is recommended that breaker points be cleaned and readjusted. However, because new points and condenser are relatively inexpensive and because a number of problems can be introduced during the cleaning operation, new points are recommended whenever the engine is dismantled to this point unless, of course, the points have been changed recently.

Sanding or filing will introduce foreign matter to the breaker point area and could prevent the points from making contact or prevent them from opening properly. Many new contact point sets are plated and the cleaning process merely removes the plating. Point life will be very short once the plating is removed.

Note that on the type illustrated above on the opposite page the movable point is opened by a removable cam on the crankshaft. If this cam is removed, as it would be to remove the crankshaft from the engine, note which way the cam came off. If the cam falls off or if you cannot remember which way it came off, consult the manufacturer's manual before replacing the cam because the flywheel, when tightened, may crack if the cam is replaced backward. Also, the points could be badly out of time if the cam is reversed.

Electronic modules that replace the points and condenser are available from some manufacturers. These solid-state units eliminate many of the problems common to breaker point ignition systems and may prove to be more reliable.

REPLACE BREAKER POINTS

Figure 24

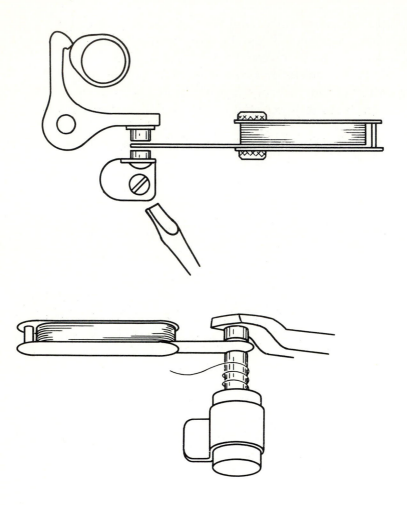

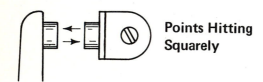 **Points Hitting Squarely**

 **Points Offset. Poor Contact**

 **Points Hitting at an Angle. Poor Contact**

Wrong

 **Points Hitting Squarely**

REPLACE BREAKER POINTS

1. Secure new breaker points and condenser. (Be sure to obtain the following information before purchasing your parts.)

Engine brand   _____

Model number   _____

Type number   _____

Serial number   _____

2. Make a sketch of the breaker points and condenser. Use the illustration on the opposite page as a sample. Carefully indicate the number of wires and the connection point of each wire.

3. Determine the correct setting of the breaker point gap. Refer to manufacturer's specifications. For engines under 7 hp, this is usually .020 inches (.51 mm). The correct gap for this engine is _____ .

4. Check the condition of the old points.

5. Check the gap of the old points. The old points were set at ____ .

6. Remove and replace the points and condenser. Place a tiny amount of lubricant on the breaker cam.

7. Align the new breaker points so that they come together squarely and without overhang. Usually the stationary contact can be adjusted. If necessary, use pliers to bend to the correct alignment. Turn the crankshaft several times to make sure that the alignment is correct.

8. Adjust the gap. Turn the crankshaft until the points are open maximum. Use the correct feeler gauge as shown at left and move the stationary contact until only a light drag is felt as the feeler gauge is moved through the contact area. Remove the feeler gauge and recheck to make sure that the points are not being pushed open as the feeler gauge enters the contact area.   **Check Point** _____

9. Clean the new contacts. Open the points and insert a small piece of hard, white, lint-free paper. Move the paper back and forth through the contact area to remove any dust or oil film from the contact area. Be sure to open the points before removing the paper so that no fragments of the paper will be torn off in the contact area.

10. Recheck wiring.   **Check Point** _____

11. Replace the breaker point box cover. Use a small amount of sealer such as Permatex to seal around wires and where the cover seats. The life of the points depends on a clean atmosphere.

Figure 25

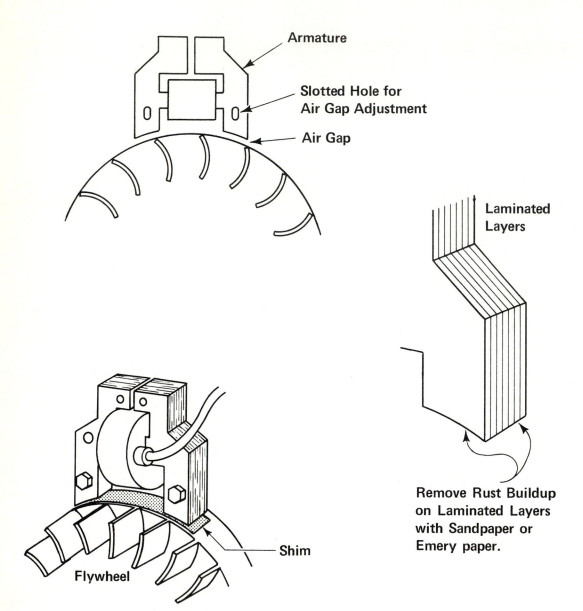

Armature

Slotted Hole for
Air Gap Adjustment

Air Gap

Laminated
Layers

Remove Rust Buildup
on Laminated Layers
with Sandpaper or
Emery paper.

Shim

Flywheel

ADJUST ARMATURE AIR GAP

On an engine whose armature is mounted outside the flywheel, as shown in Figure 25, the air gap between the armature and the flywheel must be adjusted properly.

REMOVE THE ARMATURE FROM THE ENGINE
Remove the screws holding the armature-coil assembly to the engine. Use the correct size ignition wrench or nut driver to remove the screws. Do not use pliers because the screw heads may be damaged.

CLEAN THE COOLING FINS BEHIND THE ARMATURE
If the engine is not going to be dismantled further, clean the cooling fins on the engine block behind the armature before replacing the armature.

CLEAN THE ARMATURE
Use a piece of sandpaper or emery paper to clean the armature laminations. Rust will bridge the gap between the laminations and reduce the coil efficiency. Sand all of the exposed lamination area. Coat the sanded area with a plastic protection spray or light film of grease.

REPLACE AND ADJUST THE ARMATURE
Mount the armature on the engine block. Raise the armature as high as possible and tighten one screw to hold it. Secure a piece of shim stock of the correct thickness. Check the manufacturer's specifications for the correct air gap. Usually this gap will be set approximately .010 inch (.25 mm). Hard paper such as magazine covers may be used. Add layers to get the correct thickness. See the section on measuring devices for micrometer operation to measure the thickness of the shim. Do not use a feeler gauge to set the armature air gap. Loosen the screw holding the armature and lower the armature against the shim. Tighten all screws and remove the shim.

On most smaller models, engines having the armature mounted outside the flywheel do not require timing. The only adjustment for timing is at the breaker points. Make sure that the breaker points are set correctly on this type engine.

IGNITION TIMING

Figure 26

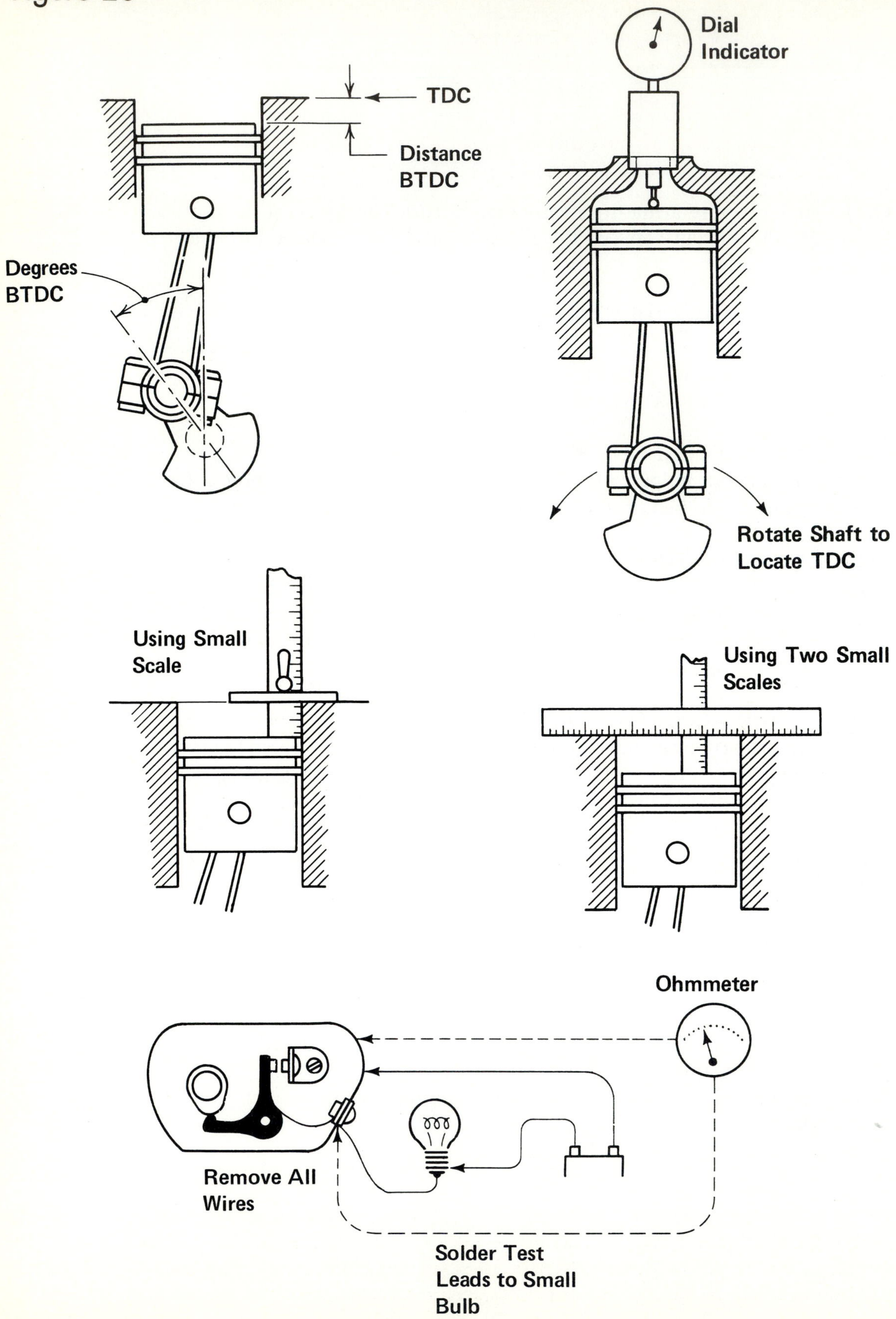

TDC

Distance
BTDC

Degrees
BTDC

Dial
Indicator

Rotate Shaft to
Locate TDC

Using Small
Scale

Using Two Small
Scales

Ohmmeter

Remove All
Wires

Solder Test
Leads to Small
Bulb

IGNITION TIMING

Engines having the armature-coil assembly mounted under the flywheel usually must have the ignition timing adjusted. The various manufacturers prescribe different techniques for the exact timing. Some types have timing marks on the armature and on the engine block. If these are used, it is a simple matter to align the marks and tighten the armature down. On this type, as with the externally mounted armature, the timing accuracy depends on the correct breaker point setting to fire the spark plug at exactly the right time. Adjust the breaker points carefully.

Other engines rely upon finding top dead center (TDC) of piston travel and then moving the piston a specified distance before top dead center (BTDC) to locate the exact point in piston travel that the points should open.

FIND TDC

To find TDC, locate the point at which the piston stops moving up and starts moving down. Study the upper illustration on the opposite page. Note that as the crankshaft passed TDC, there is a short distance of crankshaft travel that results in no piston movement. On most engines this point can be found by removing the spark plug and inserting a dial indicator or small machinist's rule. If the cylinder head is removed, the machinist's rule may be used from the top of the block. TDC is halfway between the point the piston stops moving up and begins moving downward.

CONSULT MANUFACTURER'S SPECIFICATIONS

Some manufacturers give a distance before top dead center in terms of piston travel. This distance is usually given in thousandths of an inch, such as .060 (1.53 mm) BTDC. Turn the crankshaft backward until the piston moves the specified distance.

Check Point _____

Other manufacturers give a number of crankshaft degrees before top dead center. After TDC is located, the crankshaft must be turned backward the specified number of degrees. A protractor or degree indicating wheel may be used to determine the number of degrees. If the flywheel fins are evenly spaced, as they are on some engines, count the number of cooling fins and divide into 360° to determine the number of degrees between each cooling fin. Replace the flywheel carefully and turn the crankshaft backward the specified number of degrees and remove the flywheel. Be *very* careful not to move the crankshaft while removing the flywheel.

DETERMINE WHEN THE POINTS OPEN

After the crankshaft has been brought to the correct number of degrees BTDC or thousandths of an inch BTDC, the armature–breaker point assembly is adjusted until the points are just opening. Exactly when the points break electrical contact can be determined by using an ohmmeter as shown on the opposite page. All wires must be disconnected for this test. A test lamp and battery may be used instead of the ohmmeter.

PARTS OF
THE ELECTRONIC IGNITION

Figure 27

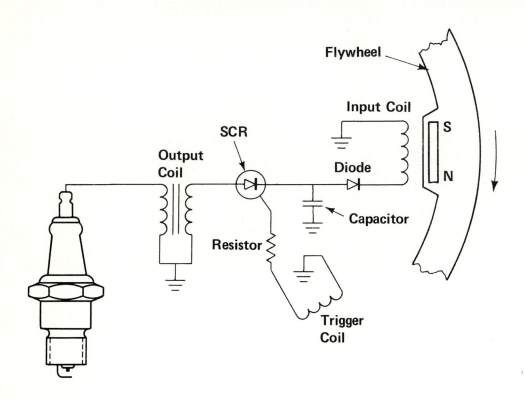

Flywheel

Input Coil

SCR

Output
Coil

Diode

S

N

Capacitor

Resistor

Trigger
Coil

SCR

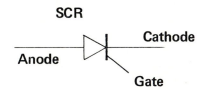

Anode

Cathode

Gate

**AN ELECTRONIC SWITCH
TURNED ON BY THE OUTPUT
OF THE TRIGGER COIL**

PARTS OF
THE ELECTRONIC IGNITION

The INPUT COIL is similar to the primary winding in the conventional armature. As the flywheel magnet passes the input coil, current flow is induced in the coil windings. The current will charge the capacitor.

The CAPACITOR is a device that will absorb and store electrical energy for a short time. When the flywheel magnet passes the input coil, the capacitor will become charged.

A DIODE is an electrical check valve. It allows current to flow in only one direction. In this circuit it allows current to flow to the capacitor, charging the capacitor. The electrical energy is then trapped in the capacitor because the diode will not allow it to go back through the coil.

The SILICON CONTROLLED RECTIFIER (SCR) is an electronic switch. It has no moving parts but normally acts as an open switch to electrical current. The switch (SCR) is turned ON by a small voltage applied to the *gate* connection.

The TRIGGER COIL provides the gate voltage to turn ON the SCR. After the magnet passes the input coil that charges the capacitor, it then passes the trigger coil that turns on the SCR, allowing the capacitor to discharge through the transformer primary winding. The trigger coil is the timing device, as the signal (voltage) to fire (turn on) the SCR is provided by the trigger coil.

The PULSE TRANSFORMER is similar to the coil in that it has a primary winding that will build up a strong, concentrated magnetic field when current passes through it and a secondary winding made up of several thousand turns of wire that will create the high voltage needed to fire the spark plug. The pulse transformer primary gets its current flow from the capacitor rather than directly from the flywheel magnet as in the flywheel magneto.

Figure 28

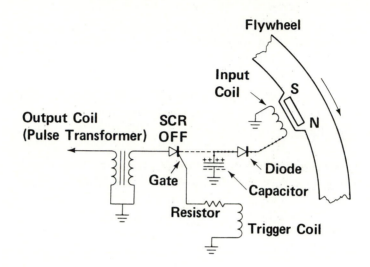

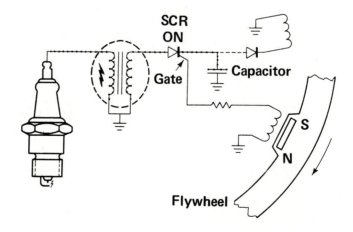

ELECTRONIC IGNITION

The solid-state electronic ignition system is used on many small gas engines. A battery is not necessary to operate the solid-state ignition because it is self-energized by the flywheel magnet as is the conventional magneto ignition system. The breaker points have been replaced with an electronic switch (SCR), which is triggered by a small coil. With this system, the only moving part is the flywheel magnet, which does not actually make moving contact with other parts.

CHARGING THE CAPACITOR
As the flywheel magnet passes the input coil, alternating current is induced in the coil. The diode, which is a one-way electrical check valve, allows the current to flow in only one direction. The flow charges the capacitor as the SCR is still turned OFF. The charge placed on the capacitor is trapped there by the diode as it is a one-way device and will not allow the current to flow back to the input coil.

TURNING ON THE SCR
After the flywheel magnet passes the input coil charging the capacitor, it then passes a small trigger coil that produces enough voltage to trigger (turn on) the SCR. A resistor is necessary to prevent excessive current flow that would damage the SCR.

DISCHARGING THE CAPACITOR
When the trigger coil turns on the SCR, it opens a path for the capacitor, which was left holding a charge, to discharge through the primary of the pulse transformer to the engine ground. This sudden burst or pulse of current through the primary winding of the pulse transformer sets up a magnetic field that induces a very high voltage in the secondary winding to fire the spark plug. The action of the primary and secondary winding in the pulse transformer is similar to that of the conventional coil. Spark timing is determined by the placement of the trigger coil.

Figure 28 (cont.)

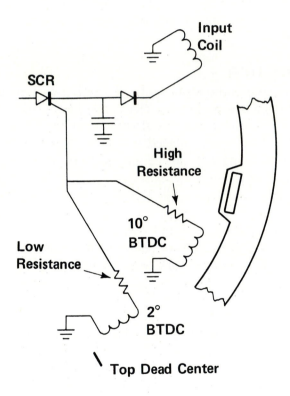

ELECTRONIC IGNITION *(cont.)*

TIMING ADVANCE

Two trigger coils could be used to trigger the SCR. The first trigger coil would be connected to the SCR through a high-value resistor that would allow this coil to fire the SCR only at high speed. The other trigger coil placed after the first would fire the SCR at low speed.

With this arrangement, the SCR would not be fired until near TDC at low speed. When the engine speed increases to the point at which the first trigger coil produces enough voltage to overcome the higher resistance in its circuit, the SCR would be fired earlier, causing earlier firing of the spark plug.

WHY ADVANCE TIMING

From the instant the spark plug fires until the burning gas mixture expands to create maximum push against the piston takes time. The time needed to "build the fire and get it going" is nearly the same at all engine speeds.

At idle speed only a few degrees of crankshaft rotation are needed to provide this time. As the crankshaft turns faster, the spark plug must fire earlier in order to provide the same amount of time to get the fuel mixture burning. This is called *timing advance*.

SERVICING THE ELECTRONIC IGNITION SYSTEM

If the unit fails to produce a good spark across the gap of a new spark plug as described on page 41, check all electrical connections to make certain that they are clean and tight. Check carefully the routing of the spark plug wire to make sure it is not damaged or grounded. If no problem is found, have the electronic unit tested by your dealer or remove the unit and replace it with a new one.

T F 1. Current flowing in a wire causes a magnetic field to surround the wire. Page 55.

T F 2. The spark plug should be cleaned periodically with an abrasive-type cleaner. Page 57.

T F 3. The flywheel magnet can be checked with an unmagnetized screwdriver. Page 67.

T F 4. A battery must be used with electronic ignition. Page 79.

T F 5. A spark plug can be used to check for secondary voltage output. Page 81.

T F 6. New spark plugs are properly set at the factory. Page 57.

T F 7. Voltage is necessary to cause current to jump the spark plug gap. Page 53.

T F 8. Spark voltage is created in the primary windings of the coil. Page 63.

T F 9. One lead from the coil primary winding is connected to ground at all times. Page 61.

T F 10. Current flows in the primary circuit while the breaker points are closed. Page 61.

T F 11. Current flows in the secondary circuit when the breaker points open. Page 63.

T F 12. The spark jumps the gap just after the piston reaches the top of its stroke. Page 63.

T F 13. The flywheel may be pried off with a screwdriver. Page 65.

T F 14. A small amount of lubricant should be placed on the cam when installing new breaker points. Page 71.

T F 15. Armature laminations should be cleaned with sandpaper to remove rust that would bridge the laminations and reduce coil efficiency. Page 73.

T F 16. It is not good practice to use pliers to remove the small armature screws. Page 73.

T F 17. A small test lamp and battery can be used instead of an ohmmeter for determining when the breaker points open. Page 75.

T F 18. A diode is like an electrical check valve. Page 77.

T F 19. Breaker points are not needed on the electronic ignition. Page 79.

20. Describe the procedure for removing a spark plug. Page 57.

21. Why are wire feeler gauges better for gapping spark plugs than the flat type? Page 57.

22. What is the purpose of the ignition system? Page 59.

23. What is the source of power used to create spark on the magneto ignition? Page 59.

24. What component in the breaker point ignition system determines the exact instant the spark will occur? Page 59.

25. The amount of current flowing in a circuit is called _____ . Page 53.

26. Pressure in an electrical circuit is called _____ . Page 53.

27. Wrapping several turns of wire on a coil would make the magnetic field _____ . Page 55.

28. Describe the sequence of events in the magneto ignition required to produce a spark. Page 59.

29. The voltage produced by the magneto may be as high as _____ volts. Page 63.

30. Describe how the flywheel is attached to the crankshaft. Page 65.

31. Why should you never strike the crankshaft end directly with a hammer? Page 65.

32. Flywheel magnets seldom loose their magnetic strength. List some conditions that might cause the magnet to be destroyed. Page 67.

33. Why do some engines use soft flywheel keys? Page 67.

34. Explain why an incorrect setting of the breaker points would affect engine timing. Page 75.

35. What information will you need to order new breaker points? Page 71.

36. Describe why new breaker points should be cleaned and how this is performed. Page 71.

37. Describe how you can get a piston at exactly top dead center. Page 75.

38. How can a protractor be used to set timing? Page 75.

39. On the electronic ignition, the pulse transformer replaces the _____ . Page 77.

40. The SCR replaces the _____ on the magneto ignition system. Page 77.

41. Why is timing advance used on an engine? Page 81.

FUEL FEED, CARBURETION, AND SPEED CONTROLS

Figure 29

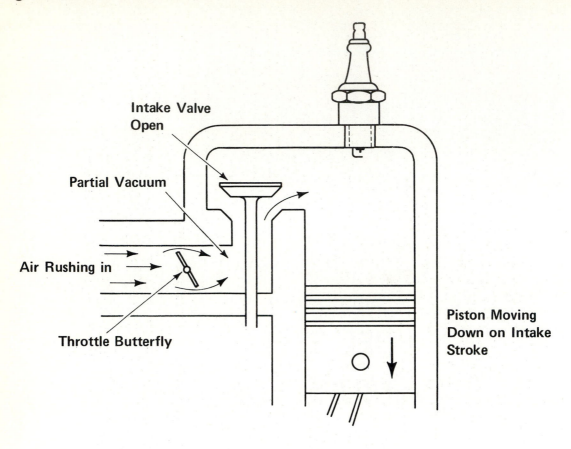

Intake Valve
Open

Partial Vacuum

Air Rushing in

Throttle Butterfly

Piston Moving
Down on Intake
Stroke

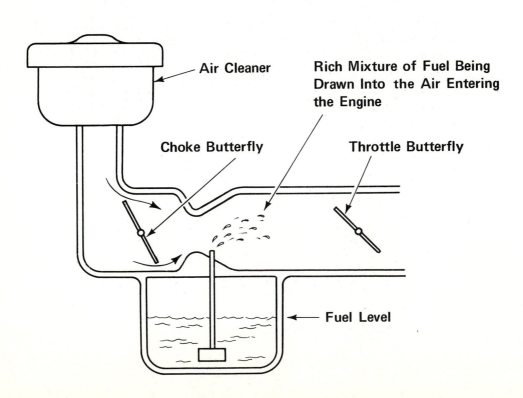

Air Cleaner

Rich Mixture of Fuel Being
Drawn Into the Air Entering
the Engine

Choke Butterfly

Throttle Butterfly

Fuel Level

CARBURETION

The carburetor controls the engine speed and provides it with a correct air-fuel mixture. The problem of providing a correct air-fuel mixture is complicated by the fact that the carburetor must provide this mixture through a wide range of engine speeds, varying loads, and varying temperatures. A carburetor that can do all of these things is complicated and expensive. On small gas engines, carburetors are designed to do as many of these things as possible while keeping the carburetor simple and inexpensive.

On four-stroke cycle engines, when the piston moves downward on the intake stroke, the intake valve is held open. Air rushes through the carburetor to fill the space being created by the moving piston. The moving piston is creating a partial vacuum above it. In this sense the engine is an air pump that pumps in large amounts of air on the intake stroke and emits large amounts of hot air and exhaust gas during the exhaust stroke.

SPEED CONTROL

The engine speed is controlled by a flat disc called the *throttle butterfly* that restricts the flow of air through the carburetor and controls the air-fuel mixture to the engine. Note that unless the throttle butterfly is wide open, the engine cannot get all the air it wants. This restriction by the throttle causes a partial vacuum to be created behind it. This is referred to as the *intake vacuum*. When the engine is running smoothly at low speed and under no load, the intake vacuum will be quite high because the throttle is restricting air flow to the engine. Opening the throttle will allow more of the air-fuel mixture to the engine and the speed will increase. An engine that is pulling hard will have the throttle open wide in proportion to the engine speed and the intake vacuum will be low. When an engine is under a light load, the intake vacuum will be high because the throttle will be limiting the air flow to the engine and the engine speed will be controlled.

THE CHOKE

The *choke* is a restriction in the outer end of the carburetor to provide an extra rich air-fuel mixture for starting. When the engine is cold, the fuel will not vaporize properly and an additional amount of fuel is needed to start the engine. The choke may be either a sliding tube or butterfly. It creates a restriction that cuts down the air flow to the engine and causes additional fuel to be drawn into the air stream. Some engines utilize a small primer pump instead of a choke. The primer pump injects a small amount of extra fuel into the air stream to create the rich mixture needed for starting.

CAUTION

Do not attempt to choke engines by putting the hand over the carburetor inlet. A misfire (backfire) could cause serious burns. Never look directly into the carburetor while the engine is being cranked.

AIR-FUEL MIXTURE

Figure 30

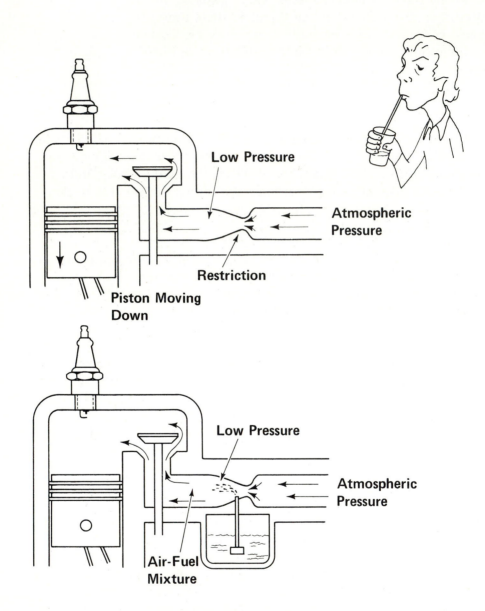

Low Pressure

Atmospheric Pressure

Restriction

Piston Moving Down

Low Pressure

Atmospheric Pressure

Air-Fuel Mixture

AIR-FUEL MIXTURE

Good combustion requires a correct air-fuel mixture. A correct air-fuel mixture is one that has enough air to completely burn all the fuel. A mixture that has too much fuel is called a *rich mixture*. An engine running on a rich mixture will have reduced power and will emit black smoke from the exhaust pipe. Running an engine with a rich air-fuel mixture will result in rapid build up of carbon on the piston, valves, and head. A mixture that does not have enough fuel for the air entering the engine is called a *lean mixture*. A lean mixture results in severe loss of power and a tendency for the engine to "surge" (speed up and slow down constantly). If the engine is operated on a lean fuel mixture under load, overheating will result, which can lead to a breakdown of the lubrication on the cylinder walls and allow the rings to scrape the cylinder walls and may cause them to seize (freeze up).

THE VENTURI

As air rushes through a restriction, it must speed up. The increased speed of the air through the restriction causes a low-pressure area in the restriction. A restriction is placed in the carburetor air horn to cause a low-pressure area. This restriction is called the VENTURI. A tube is connected from the fuel bowl into the venturi so that as air passes through the venturi, fuel is drawn into the air stream by the low pressure just as soda is drawn through a straw. Creating a low pressure in your mouth causes the soda to be drawn into your mouth. When the throttle is opened wider, the air flow will increase and the pressure in the venturi will become even lower, increasing the fuel flow into the airstream. The amount of fuel drawn out of the fuel bowl is determined by the amount of air flowing through the venturi. The mixture is consistent over a wide range of speeds. Opening the throttle wider will allow more air through the carburetor. This additional air will cause more vacuum in the venturi, which will draw additional fuel into the airstream. The proportion of air to fuel will be the same, providing enough air to properly burn the fuel allowed into the cylinder. Also, mixing fuel with the air in the venturi helps to turn the fuel into a vapor mist rather than drops of fuel. Fuel that reaches the cylinder in droplet form will not burn properly. Only fuel that is changed into a vapor will burn properly in the engine. A rich fuel mixture allows drops of fuel into the engine. Instead of burning quickly like the vapor, the fuel drops tend to "wash" the oil film from the cylinder wall and cause undue wear on the rings and cylinder wall.

Figure 31

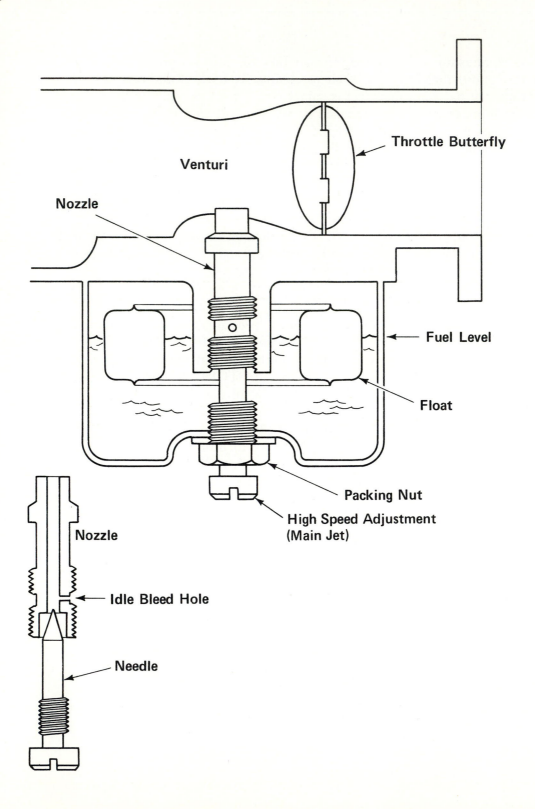

Throttle Butterfly

Venturi

Nozzle

Fuel Level

Float

Packing Nut

High Speed Adjustment
(Main Jet)

Nozzle

Idle Bleed Hole

Needle

THE HIGH-SPEED AND
INTERMEDIATE-SPEED CIRCUIT

The high-speed or load circuit is the real heart of the carburetor, for fuel economy, engine power, engine temperature, and engine life are all dependent on the correct mixture during normal speed and load operation. The illustration at the left shows the major components of the circuit during normal operating conditions. The throttle butterfly is regulating the air flow into the engine. The venturi is creating a low-pressure area to draw the fuel from the fuel bowl. The carburetor shown here utilizes a float to maintain a constant fuel level. Some carburetors use other methods to maintain a constant fuel level. The fuel level is very important. An increase of as little as one-sixteenth inch in the fuel level will make the fuel easier to pull into the airstream in the venturi and will cause a rich mixture. A low level will cause a lean air-fuel mixture.

The *nozzle* is a precision tube that usually is removable. It provides the seat for the needle valve and extends into the venturi as shown in the illustration. The nozzle is made of brass or other soft material and is subject to wear and damage from abuse.

The high-speed fuel metering adjustment needle valve adjusts the flow of fuel into the nozzle. It limits the amount of fuel that will be drawn into the airstream flowing through the venturi. This adjustment is made with the engine warm and running at least at half speed. Never operate engines at high speed when the engine is not under load! The engine may reach a speed at which mechanical failure can occur in the connecting rod or other fast-moving parts. NOTE: The high-speed jet is "fixed" on some engines. On these engines the high-speed fuel flow cannot be adjusted.

Notice the *idle bleed hole* located in a flat ring turned in the threaded portion of the nozzle. This allows fuel to enter the idle circuit (see Idle Circuit that follows) on this type carburetor. On some older models this area of the threads was not turned flat but was threaded all the way up. On these models the old nozzle cannot be replaced in the carburetor because the idle bleed hole will not align with the matching hole in the carburetor. If these holes do not align, there will be a loss of the idle circuit and the engine will die when the throttle is returned to the low speed position. Once the nozzle has been moved for any reason, it should be completely removed. If it does not have a flat area completely around the nozzle in the thread area, it must be replaced with a new nozzle.

THE IDLE CIRCUIT

Figure 32

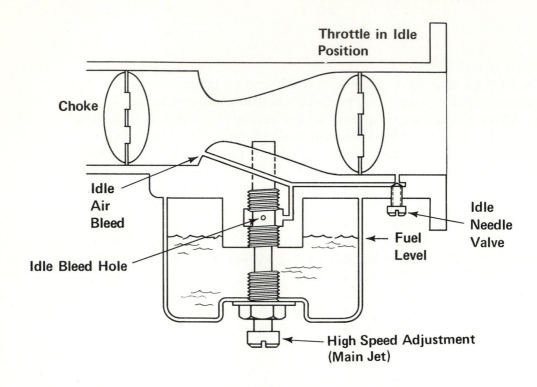

Throttle in Idle Position

Choke

Idle Air Bleed

Idle Bleed Hole

Fuel Level

Idle Needle Valve

High Speed Adjustment (Main Jet)

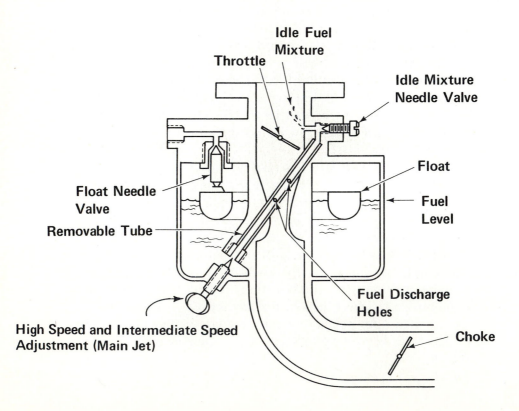

Idle Fuel Mixture

Throttle

Idle Mixture Needle Valve

Float Needle Valve

Removable Tube

Float

Fuel Level

High Speed and Intermediate Speed Adjustment (Main Jet)

Fuel Discharge Holes

Choke

THE IDLE CIRCUIT

The *idle circuit* is necessary to supply a correct air-fuel mixture to the engine when the engine is at idle and low speeds. Air flow through the venturi at low speeds is too slow to create the low-pressure area needed to draw the fuel into the venturi. A low-pressure area (vacuum) exists behind the throttle butterfly during idle because the throttle is restricting the air flow into the engine. A separate idle circuit connected behind the throttle comes into operation when a high vacuum exists behind the throttle.

Gas is drawn by the low pressure behind the throttle butterfly from the fuel bowl past the idle mixture screw. Air is also drawn into the idle circuit through the air bleed in the air horn near the venturi. Some air is also entering the engine around the throttle butterfly. This air has no fuel for it passed through the venturi too slowly to draw fuel from the fuel bowl. This air entering around the throttle and the air carrying a rich mixture of fuel is entering through the idle mixture needle valve. Adjusting the idle mixture screw limits the amount of air fuel entering from the idle circuit, which mixes with the air entering around the throttle butterfly. This adjustment is made only with the engine at operating temperature and at the correct idle speed (see Basic Carburetor Adjustment).

Some carburetors utilize the fuel discharge holes in the venturi for air bleed during idle (see the illustration at the bottom of the opposite page). During high-speed operation the low pressure in the venturi draws fuel through the high-speed fuel needle valve and through the fuel discharge holes in the tube extending through the venturi. When the throttle is closed and the vacuum is greater behind the throttle butterfly than in the venturi, air enters through the fuel discharge holes on its way through the idle circuit. Note that during idle the fuel also comes through the high-speed needle valve. The adjustment of high-speed needle valve can also affect the idle mixture. Always adjust the high-speed mixture first and remember that anytime the high-speed mixture is changed, the idle mixture should be readjusted.

CARBURETOR FUEL SUPPLY

Figure 33

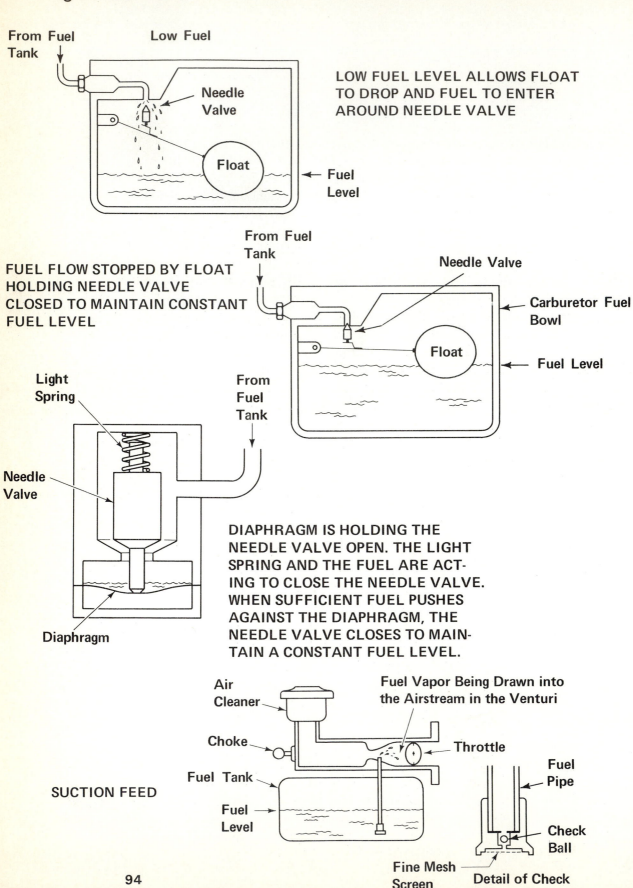

From Fuel Tank

Low Fuel

Needle Valve

Float

Fuel Level

LOW FUEL LEVEL ALLOWS FLOAT TO DROP AND FUEL TO ENTER AROUND NEEDLE VALVE

FUEL FLOW STOPPED BY FLOAT HOLDING NEEDLE VALVE CLOSED TO MAINTAIN CONSTANT FUEL LEVEL

From Fuel Tank

Needle Valve

Carburetor Fuel Bowl

Float

Fuel Level

Light Spring

From Fuel Tank

Needle Valve

Diaphragm

DIAPHRAGM IS HOLDING THE NEEDLE VALVE OPEN. THE LIGHT SPRING AND THE FUEL ARE ACTING TO CLOSE THE NEEDLE VALVE. WHEN SUFFICIENT FUEL PUSHES AGAINST THE DIAPHRAGM, THE NEEDLE VALVE CLOSES TO MAINTAIN A CONSTANT FUEL LEVEL.

Air Cleaner

Fuel Vapor Being Drawn into the Airstream in the Venturi

Choke

Throttle

Fuel Tank

Fuel Pipe

SUCTION FEED

Fuel Level

Check Ball

Fine Mesh Screen

Detail of Check Ball

CARBURETOR FUEL SUPPLY

A constant fuel level below the venturi is necessary to maintain a stable air-fuel mixture. Slight differences in the fuel level in the carburetor float bowl will affect engine power and performance. The fuel level can be controlled in a number of ways. A few of the more common techniques will be discussed here.

THE FLOAT

A common method of controlling the fuel level is by using a float and needle valve. With no fuel in the float bowl of the carburetor, the float is lowered, leaving the needle valve open. The fuel will flow from the tank into the float bowl and will raise the float. When the fuel reaches the level determined by the float setting, the needle valve closes against the seat and stops the fuel flow. As fuel is drawn out of the fuel bowl, the float drops slightly, allowing the needle valve to admit more fuel. The fuel level is very accurately maintained by this method. Float-type carburetors are limited to near-level operations. They are more expensive to build and repair, and they require very precise adjustment.

THE DIAPHRAGM CARBURETOR

This carburetor uses a light flexible diaphragm to open the needle valve to admit fuel. The diaphragm holds the needle valve open to admit fuel. The fuel enters the bowl and pushes against the diaphragm, allowing the light spring to close the valve. Thus, a constant amount of fuel is maintained by the diaphragm. This carburetor can be used on engines that must be tilted during operation, for example, outboards and chain saws. Sometimes engine vacuum is utilized in aiding the control of the diaphragm.

SUCTION FEED

The suction-feed carburetor uses a shallow fuel tank mounted directly below the venturi section of the carburetor. The shallow fuel tank replaces the float bowl. Because the tank is shallow, the difference between the fuel level in a full tank and a near empty tank will not create excessive differences in the fuel mixture. However, it does tend to provide a rich mixture when the fuel tank is full and a lean mixture when the tank is nearly empty. It is therefore necessary to have the tank half-full when adjusting the carburetor. This carburetor may not have an idle mixture adjusting screw. The idle mixture will be dependent on the high-speed mixture. This carburetor is simple and maintenance free. It requires a clean air cleaner, clean fuel, and a tight connection to the engine.

The fuel pipe has a fine mesh inlet screen and may have a check ball. If fuel has been allowed to remain in the engine during storage or if dirt has gotten into the tank, the screen and check ball may become clogged. Remove and clean thoroughly. If the fuel tube is brass, it is a press fit in the carburetor body. To remove a brass fuel tube, clamp the tube lightly in a vise and pull the carburetor off. If the tube is nylon, it is threaded into the carburetor body and is removed with a socket wrench.

BUILT-IN FUEL PUMP

Figure 34

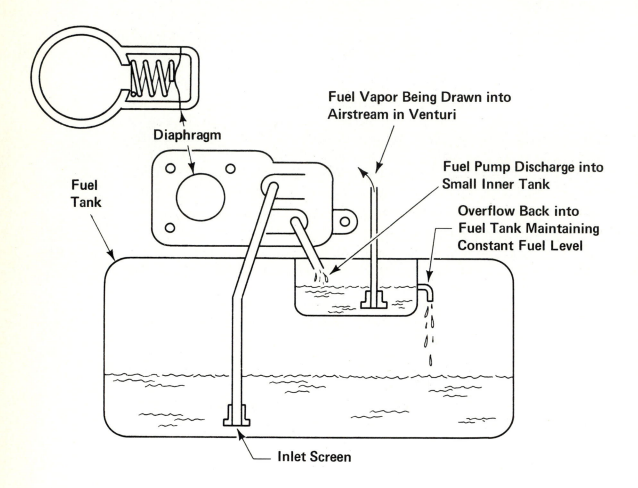

Diaphragm

Fuel Vapor Being Drawn into Airstream in Venturi

Fuel Pump Discharge into Small Inner Tank

Fuel Tank

Overflow Back into Fuel Tank Maintaining Constant Fuel Level

Inlet Screen

VACUUM–OPERATED INTEGRAL FUEL PUMP

BUILT-IN FUEL PUMP

A fuel pump built into the carburetor may be utilized to maintain a constant fuel level, which eliminates the need for the float. On single-cylinder, four-stroke cycle engines there is a partial vacuum just behind the throttle plate on the intake stroke as the piston moves downward and the throttle restricts the air flow. During the other three strokes the intake area is at normal pressure. This creates a pulsing vacuum in the intake area between the intake valve and the throttle. A flexible diaphragm placed in this area is pulled in by the intake vacuum during the intake stroke and is released during the compression, power, and exhaust strokes. The flexible diaphragm will be moving back and forth. The other side of this pulsating diaphragm can then be used as a fuel pump to draw gas from a large fuel tank up into a small inner tank. Flapper-type check valves cut into the diaphragm direct the fuel through the pump into the inner tank. The fuel pump can pump fuel faster than the engine can use it. The excess fuel is allowed to overflow back into the large fuel tank, thus maintaining a constant fuel level in the small inner tank without using a float.

The air flow through the venturi causes a partial vacuum that draws the fuel from the inner tank just as it does from the float bowl on the float-type carburetor.

Many of these models also have the magneto stop switch on the carburetor speed control. When removing or servicing this carburetor, care should be taken to note the location of the wire coming from the magneto. If this wire comes in contact with the engine ground, the magneto will be disabled.

Diaphragm failure may be the cause of hard starting or may lead to an excessively rich fuel mixture if the diaphragm ruptures. A hole in the diaphragm would allow fuel to be pulled from the fuel pump directly into the intake manifold. Diaphragm replacement is simple, but care must be taken to get the diaphragm placed correctly in the fuel pump.

ENGINE SPEED CONTROLS

Figure 35

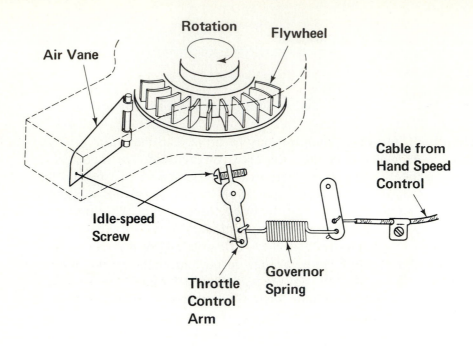

Rotation

Flywheel

Air Vane

Cable from
Hand Speed
Control

Idle-speed
Screw

Throttle
Control
Arm

Governor
Spring

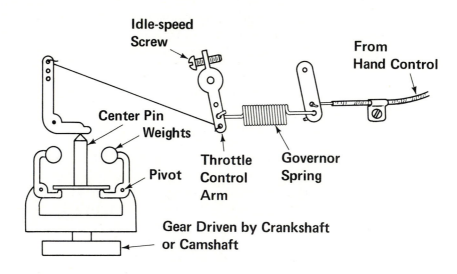

Idle-speed
Screw

From
Hand Control

Center Pin

Weights

Pivot

Throttle
Control
Arm

Governor
Spring

Gear Driven by Crankshaft
or Camshaft

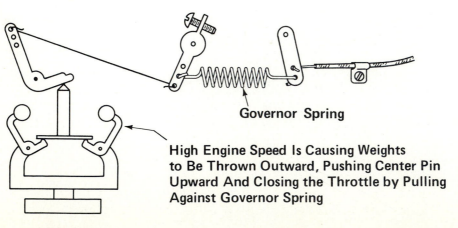

Governor Spring

High Engine Speed Is Causing Weights
to Be Thrown Outward, Pushing Center Pin
Upward And Closing the Throttle by Pulling
Against Governor Spring

ENGINE SPEED CONTROLS

Engine speed is controlled by the operator's manual speed control and the governor. Most small gasoline engines used to power a constant load such as a mower or tiller must employ a governor to maintain a nearly constant speed as the load changes. Changing or removing the linkages on lawnmowers usually results in the loss of the governor as a speed control. The engine speed is then nearly uncontrollable. The engine will act as though it has almost no power and it will slow down excessively upon encountering a load. The operator must then increase the engine speed with the manual speed control lever. Once the load is reduced, the engine will overspeed. A normally functioning governor is essential for satisfactory operation on machines such as mowers, tillers, and generators.

AIR VANE GOVERNORS

The *air vane governor* is a simple and inexpensive device for controlling engine speed. An air vane located under the shrouding near the flywheel senses the engine speed by the amount of wind created by the fins of the spinning flywheel. As the engine speed increases, the increased force of the air flow created by the flywheel will push against the air vane. The air vane is connected by a wire linkage to the throttle control arm. It pulls against the governor spring to close the throttle, thereby reducing the engine speed. The speed of the engine is determined by a balance between the pull created by the air valve and the pull on the governor spring from the hand control.

If the operator moves the hand control to increase the engine speed, the governor spring will pull harder on the throttle control arm opening the throttle wider. The engine speed will increase until the air flow from the flywheel against the air vane increases to equal the pull of the governor spring.

If the engine encounters an increase in load, the engine speed begins to decrease and the air flow against the air vane will decrease, allowing the governor spring to pull the throttle open wider to retain the engine speed. The governor will maintain constant speed as conditions vary.

MECHANICAL GOVERNORS

Mechanical governors are considered to be somewhat more precise and dependable than air vane governors. They are more expensive and add weight to the engine. The mechanical governor senses the engine speed by the amount of centrifugal force exerted on rotating weights located inside the crankcase. The type shown in the accompanying illustration has weights mounted on a pivot so that as the weights are thrown outward by centrifugal force the center pin is lifted upward. As the engine speed increases, the weights are spinning faster and the force exerted by the weights to the center pin is increased. This acts to slow down the engine speed. Again, as with the air vane governor, the engine speed is determined by the balance of force created on the governor spring by the hand speed control and the force produced by the spinning weights. If a load causes a reduction in the engine speed, the centrifugal force acting on the spinning weights will be reduced, allowing the governor spring to open the throttle to increase the engine speed. As the engine speed recovers, the governor will again exert pull against the governor spring to maintain the engine speed.

Figure 36

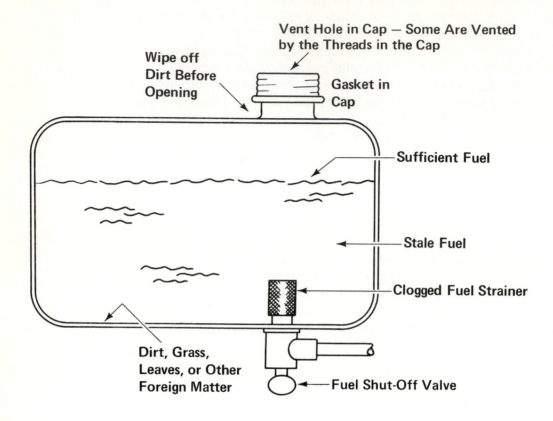

Vent Hole in Cap — Some Are Vented
by the Threads in the Cap

Wipe off
Dirt Before
Opening

Gasket in
Cap

Sufficient Fuel

Stale Fuel

Clogged Fuel Strainer

Dirt, Grass,
Leaves, or Other
Foreign Matter

Fuel Shut-Off Valve

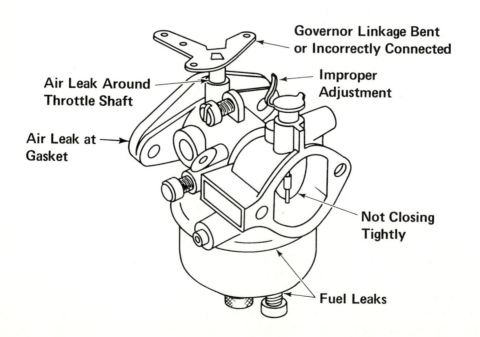

Governor Linkage Bent
or Incorrectly Connected

Air Leak Around
Throttle Shaft

Improper
Adjustment

Air Leak at
Gasket

Not Closing
Tightly

Fuel Leaks

FUEL SYSTEM SERVICE

Many of the problems commonly associated with the fuel system may actually be electrical problems such as a bad spark plug or a weak spark caused by a bad condenser or breaker points. Even a rusted magneto armature may cause symptoms similar to fuel problems. Check the condition of the spark plug (see page 6, Spark Plug Diagnosis) and the quality of the spark before blaming the fuel system for failure to start or for poor performance. If the spark plug condition is good and if a strong spark is present, the following suggestions may help find the problem.

ENGINE FAILS TO START

Check Fuel Tank. Make sure that there is sufficient fuel. Make certain, too, that the fuel is fresh. Stale gasoline can cause hard starting and poor performance. Check the fuel tank for dirt, grass, etc. Mowers and tillers tend to collect foreign matter in the fuel tank. Remove the tank and clean it completely if foreign matter is present. It is good practice to wipe off the top of the fuel tank before removing the filler cap.

Check the Choke. On many small gasoline engines the choke is operated automatically by the hand speed control lever. Improper adjustment of the speed control cable may be causing the choke to close completely. If necessary, remove the air cleaner to determine if the choke is closing completely.

Remove the Spark Plug. After cranking the engine several times with the choke fully closed, remove the spark plug and observe the condition of the plug (see page 6, Spark Plug Diagnosis).

ENGINE STARTS BUT SURGES INSTEAD OF RUNNING SMOOTHLY

Incorrect Fuel Mixture. Adjust the carburetor (see Carburetor Adjustment section). Surging usually indicates a lean fuel mixture unless accompanied by black smoke from the exhaust pipe.

Incorrect Governor Operation. Incorrect connection of governor linkages, bent or broken linkages, incorrect linkages, or binding or rubbing of the linkages can cause incorrect governor operation. Check these linkages for free operation. On mechanical governors, check for correct adjustment.

Air Leaks. Air leaks into the carburetor can occur at the gasket where the carburetor attaches to the engine. Most carburetors are held on by two capscrews that are readily loosened by vibration, allowing air to enter past the gasket. Do not overtighten these capscrews for they are easily stripped in the aluminum engine block. Also check the throttle shaft for air leaks. If the throttle shaft is loose in its bushings, air will enter around the throttle shaft and will cause the fuel mixture and the engine speed to change as the throttle shaft moves around in the bushing. This condition usually calls for replacement of the carburetor body.

Figure 37

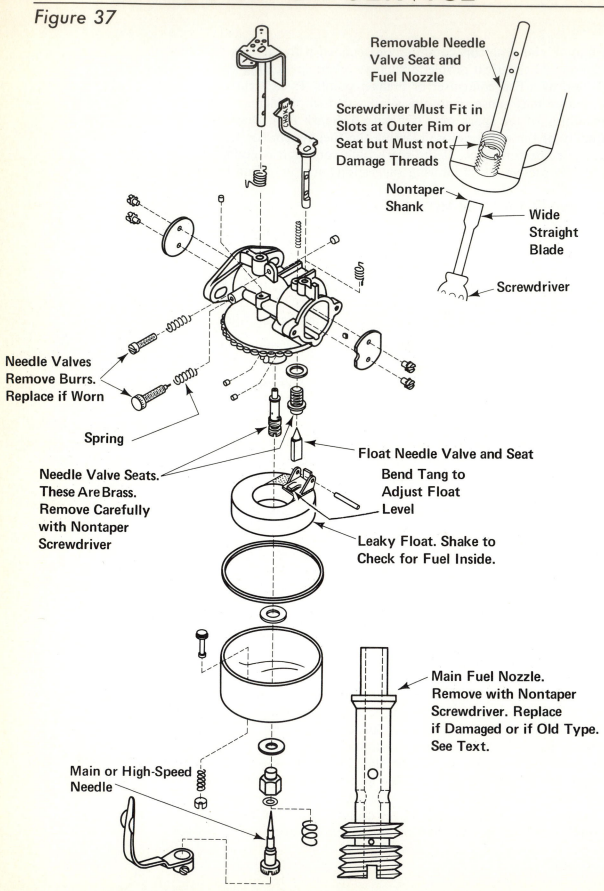

Removable Needle Valve Seat and Fuel Nozzle

Screwdriver Must Fit in Slots at Outer Rim or Seat but Must not Damage Threads

Nontaper Shank

Wide Straight Blade

Screwdriver

Needle Valves Remove Burrs. Replace if Worn

Spring

Needle Valve Seats. These Are Brass. Remove Carefully with Nontaper Screwdriver

Float Needle Valve and Seat

Bend Tang to Adjust Float Level

Leaky Float. Shake to Check for Fuel Inside.

Main Fuel Nozzle. Remove with Nontaper Screwdriver. Replace if Damaged or if Old Type. See Text.

Main or High-Speed Needle

CARBURETOR SERVICE

In most cases it will be necessary to remove the carburetor if disassembly is required. The first step is to draw a diagram of the speed control linkages and springs because the proper operation depends on the correct connection of these linkages. If these linkages are already disconnected or missing, it will be necessary to consult the manufacturer's literature or find a similar engine to determine the correct connection. Do not bend the linkages when removing them from the carburetor. Bending the linkages will change the length and affect governor operation. Remove the bolts that hold the carburetor to the engine and twist the carburetor assembly to disconnect the linkage. Needle nose pliers are helpful in removing springs. Be very careful not to stretch the springs.

An assortment of good screwdrivers, pliers, and a clean, well-lighted work area can be very helpful for carburetor disassembly, cleaning, and reassembly. Clean the exterior of the carburetor with a stiff brush and cleaning solution before disassembly. If the carburetor is to be disassembled, it is usually worthwhile to rebuild the carburetor. A kit can be purchased that contains the necessary parts that normally wear and cause trouble, for example, gaskets, needle valves, and seats. Check the kit to determine the parts that are provided before disassembly to determine which parts should be removed.

After disassembly, the carburetor should be soaked for a few minutes in carburetor cleaner to remove the gum and varnish residue. *Caution:* Carburetor cleaner is extremely painful if it is brought in contact with the eyes or skin. Avoid contact with the cleaner. Wash the parts in water thoroughly after cleaning them with carburetor cleaner. Do not attempt to brush or blow out the carburetor passages until they are thoroughly washed out with water. Remove all gaskets and plastic parts. Do not place them in the carburetor cleaner, because some of them will be dissolved by the cleaner.

Remove and replace needle valves and seats very carefully. They are made of soft material and can be very easily damaged. Use only screwdrivers that are flat across the tip and are the correct width for the screw. Use wrenches instead of pliers to remove fuel lines and hexagon head capscrews. When removing needle valve seats that are recessed, use a screwdriver that is wide enough to make good contact with the seat but not so wide that it will damage the threads on the sides. Usually a nontaper screwdriver is required.

Check the float for leaks. If a hollow brass float is used, leaks will allow fuel to fill the float. Shake the float near your ear to determine if there is fuel in the float.

Before replacing the needle valve seats and the needle valves, visually inspect the fuel passages to make sure that they are clean. Reassemble the carburetor with care. Remember to reconnect the linkages and magneto stop wire (if so equipped) before attaching the carburetor.

Figure 38

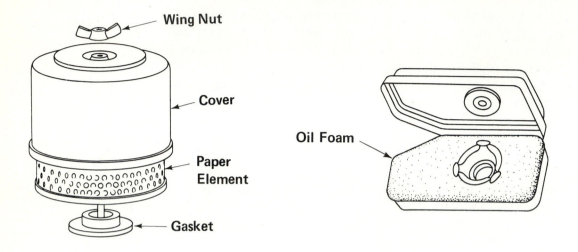

DRY PAPER TYPE FILTER   POLYURETHENE FOAM TYPE FILTER

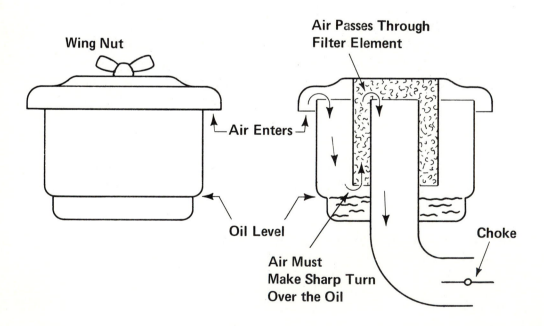

AIR CLEANER SERVICE

Small gasoline engines are often exposed to very dirty working conditions. A good, clean air filter is essential to efficient operation and long life. Remember that the engine is really quite an efficient air pump. An engine with a displacement of eight cubic inches will try to draw in eight cubic inches of air on *each* intake stroke. You can readily see that an engine running at full throttle will consume a great deal of air. An air cleaner must be kept in good condition to admit that much air without allowing dirt into the engine. If the air filter starts to become clogged with dirt, grass, or other matter, it will restrict the flow of air through the carburetor. This is exactly what the choke does. A dirty air cleaner will choke the engine, causing it to run on a rich air-fuel mixture, which will cause loss of power, spark plug fouling, and carbon buildup. The excess gas will tend to "wash down" the oil film from the cylinder walls, causing improper ring lubrication and will lead to excessive wear or possibly engine lock up. Also, the excess fuel in the mixture will get past the rings into the crankcase, diluting the oil.

DRY PAPER-TYPE FILTER
This type of air filter utilizes porous paper to allow air flow and collect dirt and foreign matter. Tap the filter element to jar the dirt loose. Blow dirt away by blowing from inside out with low pressure air. A vacuum cleaner is effective for cleaning this filter. If the filter still appears clogged, replacement is necessary. Do not attempt to soak or wash the dry paper element.

POLYURETHENE ELEMENTS
Air filters of this type can usually be cleaned several times. Clean the filter thoroughly with soap and warm water. Squeeze out and add a few drops of light oil before replacing. Make sure that the filter fits back into the metal holder and is retained so that it cannot be drawn into the engine.

OIL BATH AIR CLEANERS
Oil bath air cleaners are usually used on engines that are used regularly in dirty conditions. This air cleaner is capable of collecting larger amounts of dirt and is easily serviced. The incoming air travels downward toward the oil at a high rate of speed. The air is forced to make a sharp turn just above the oil. Dirt and foreign matter tend to travel straight into the oil and stick there. Most of the dust and dirt entering the air cleaner will collect in the oil. The air then travels upward through a filter element before entering the engine. Check this element and wash it out if necessary. If the oil is changed regularly, the air cleaner will do a very efficient job of protecting the engine from dirt. The interval for oil change is determined by the amount of use and the operating conditions. Usually, instructions are labeled on the air cleaner.

1. Describe intake vacuum under heavy load. Page 87.

2. Why is it dangerous to use your hand to choke the engine? Page 87.

3. Excessive carbon will build up inside an engine operated with a _____ fuel mixture. Page 89.

4. Describe the action in the venturi. Page 89.

5. Why is a special circuit necessary for idle speeds? Page 93.

6. Why is the idle mixture adjusted after the high-speed circuit? Page 93.

7. What conditions are necessary when adjusting the suction feed carburetor? Page 95.

8. What moves the flexible diaphragm on the carburetor fuel pump? Page 95.

9. What determines the fuel level on carburetors with a built-in fuel pump? Page 97.

10. Some engines have an electrical wire connected to the throttle speed control. What is the purpose of this wire? Page 97.

11. What is the purpose of the governor on a small gas engine? Page 99.

12. Describe an air vane governor. Page 99.

T F 13. A rich fuel mixture will cause an engine to wear rapidly. Page 89.

T F 14. A high or low float level can affect fuel mixture. Page 95.

T F 15. Most small-engine carburetors have one adjustment for air and one for fuel. Page 43.

T F 16. The throttle must be wide open and the engine under no load to adjust the high-speed circuit. Page 43.

T F 17. Stale gasoline can cause hard starting. Page 101.

T F 18. Choke linkage adjustment is a common cause of hard starting. Page 101.

T F 19. The capscrews that hold the carburetor to the engine block should be tightened very tightly. Page 101.

T F 20. A dirty air cleaner will have the effect of choking the engine. Page 105.

T F 21. An incorrectly adjusted carburetor will cause an engine to surge. Page 101.

T F 22. What is a typical idle speed for lawnmower engines? Page 43.

T F 23. The carburetor controls engine speed and provides it with _____ . Page 87.

T F 24. Engine speed is controlled by a flat disc called the _____ . Page 87.

T F 25. Describe intake vacuum at idle speed. Page 93.

MEASURING DEVICES

Section VII

ENGINE MEASUREMENTS— DISASSEMBLY INSTRUCTIONS AND PRECAUTIONS

MEASURING DEVICES

Figure 39

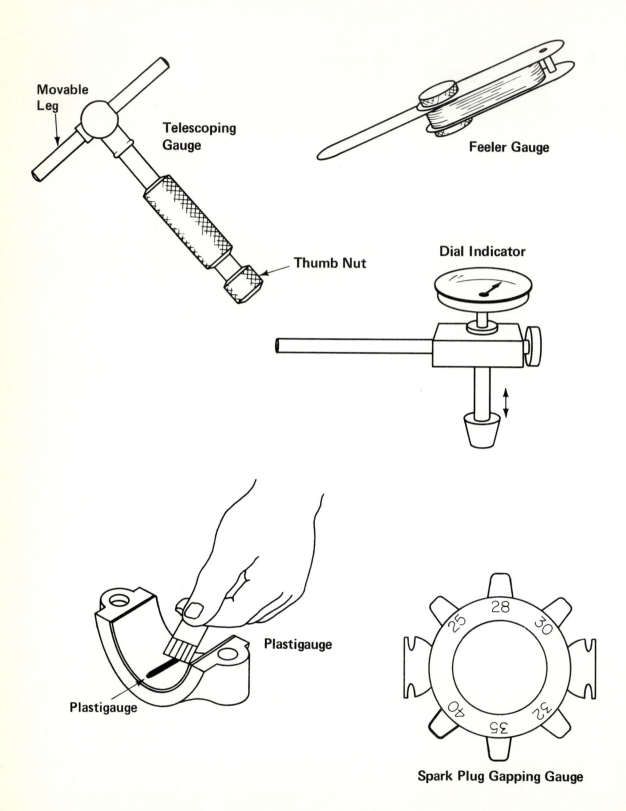

Movable Leg

Telescoping Gauge

Thumb Nut

Feeler Gauge

Dial Indicator

Plastigauge

Plastigauge

Spark Plug Gapping Gauge

25 28 30 32 35 40

MEASURING DEVICES

TELESCOPING GAUGE

The *telescoping gauge* gets its name from the way the "legs" of the device slide together. To use the telescoping gauge, first loosen the thumb nut and slide the legs together. When the legs are together, tighten the thumb nut. The gauge is then placed inside the hole or space to be measured. The thumb nut is then released. This permits the legs to spring outward against the sides of the hole. With the handle parallel to the sides of the hole, tighten the thumb nut. The gauge is then removed and the distance across the legs is measured with an outside micrometer.

FEELER GAUGE

A *feeler gauge* or thickness gauge is used to measure small distances or clearances. This tool measures in thousandths of an inch or millimeters. The flat feeler gauge contains blades (thin strips of metal) of different thicknesses. The thickness is stamped on each blade. By placing different blades together many different measurements can be made.

SPARK PLUG GAUGE

The *spark plug gauge* normally is made from small wires. The diameter of each wire (thickness) is marked on the gauge. The hook on this gauge is used to bend (adjust) the ground electrode of the spark plug.

DIAL INDICATOR

The *dial indicator* measures distances in thousandths of an inch. This device is fastened securely in place by special holders and clamps. The pad is pressed against the shaft or unit being tested for movement such that the needle gives a reading. The outer ring of the gauge is rotated until the needle is on 0. The shaft is then moved or turned. The movement of the needle indicates the variation in thousandths of an inch or millimeters.

PLASTIGAUGE

A *Plastigauge* is used to measure the clearance between a plain bearing and the journal. The clearance is determined by placing a piece of the Plastigauge (stringlike plastic) across the bearing and then assembling the bearing cap. The cap is tightened to specifications. *Without* moving the shaft, the cap retaining nuts or bolts are loosened and the cap is removed. The flattened Plastigauge will be on the shaft or the bearing. The width of the flattened Plastigauge is compared to the scale on the Plastigauge package to determine the bearing clearance. Plastigauge is available in several different sizes. Clearances up to nine thousandths (.229 mm) can be measured by this method.

THE MICROMETER

Figure 40

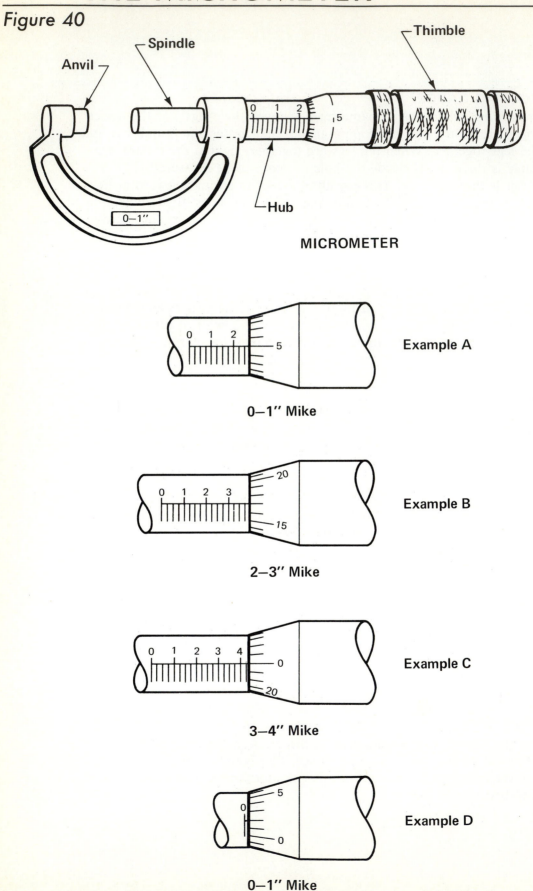

MICROMETER

Example A

0—1″ Mike

Example B

2—3″ Mike

Example C

3—4″ Mike

Example D

0—1″ Mike

THE MICROMETER

A *micrometer* measures distances in thousandths of an inch. Micrometers come in different sizes. For example, a 0–1″ "mike" will measure from 0 to 1 inches in thousandths of an inch; a 2–3″ mike will measure from 2 to 3 inches in thousandths of an inch.

One complete turn of the thimble changes the distance between the anvil and the spindle 25 thousandths (.025″). Each number on the hub is read as hundred thousandths. Each mark on the thimble is one thousandth.

To read a mike, first note the size of the micrometer—1″, 2″, etc. Second, read the longest numbered line that is visible. Third, count .025 for each of the short marks on the hub to the right of the numbered line. Fourth, to these readings add the reading on the thimble.

Example A.
1. 0.000 (0–1″ mike)
2.   .200 largest numbered line
3.   .050 two short marks beyond numbered line
4.   .005 5 on the thimble
     .255 = distance between anvil and spindle

Example B.
1. 2.000 (2–3″ mike)
2.   .300 largest numbered line
3.   .075 three short marks beyond numbered line
4.   .017 17 on the thimble
    2.392 = distance between anvil and spindle

Example C.
1. 3.000
2.   .400
3.   .025
4.   .000
    3.425 = distance between anvil and spindle

Example D.
1.   .000
2.   .000
3.   .000
4.   .002
     .002 = distance between anvil and spindle

Metric micrometers are read in a manner very similar to the English micrometer.

DISASSEMBLY AND REASSEMBLY INSTRUCTIONS AND PRECAUTIONS

Figure 41

POOR COMPRESSION

HIGH OIL COMPRESSION

ENGINE KNOCKS

ENGINE "LOCKED"

DISASSEMBLY AND REASSEMBLY INSTRUCTIONS AND PRECAUTIONS

The engine should be dismantled *only* if there is good reason. Thorough testing can determine whether or not complete teardown is necessary. (See the Compression Testing section, page 39.)

REASONS FOR DISASSEMBLY

1. Poor compression because of internal wear or leak at head gasket.
2. High oil consumption because of worn rings.
3. Excessive noise or knocks coming from inside the crankcase.
4. Crankshaft won't turn because of internal problem.
5. Excessive oil leaks around the crankshaft seals or crankshaft gasket.
6. Failure of lubrication system.

PRECAUTIONS

Safety.   Wear safety glasses when performing cleaning operations!

Take your time.   Do not be in a hurry when disassembling the engine. Observe how the unit is assembled before taking any component apart. If it is a first-time experience, it is wise to make notes and sketches of how parts go together.

Work in a clean work space.   Lay parts in order on the work bench as they come apart. Place small parts and bolts in a container. In some instances it may be helpful to thread bolts into their respective engine parts rather than to place them in a container.

NEEDED TOOLS

Disassembly of a small engine does not require extensive special tools and equipment. The following common hand tools are recommended:

3/8" square drive socket set with reversible ratchet and extensions (sizes 1/4" through 7/8" by sixteenths or 7mm through 21mm for metric)

3/8" square drive torque wrench (150 inch-pound capacity or 173 CK for metric)

Combination wrench set (5/16" through 3/4" by sixteenths or 8mm through 19mm for metric)

Ball peen hammer (12 oz)

Combination slip joint pliers

Needle nose pliers

Assorted flat blade screwdrivers (3)

Phillips screwdriver

Carbon scraper or putty knife

Flat feeler gauge

T F   1. A telescoping gauge may be used in measuring the cylinder. Page 109.

T F   2. A feeler gauge consists of several thin strips of metal of different thicknesses. Page 109.

T F   3. Plastigauge is used to measure the spark plug gap. Page 109.

T F   4. A bent crankshaft would be identified by using a telescoping gauge. Page 109.

T F   5. A micrometer measures in fractions of an inch. Page 111.

T F   6. One turn of the micrometer thimble changes the distance between the anvil and spindle 25 thousandths. Page 111.

T F   7. A one-inch micrometer is designed to measure up to 2 inches. Page 111.

T F   8. One should not disassemble the engine until good reason for a "tear down" is established. Page 111.

T F   9. Extensive special tools are needed for disassembling the engine. Page 111.

T F 10. It is helpful to make notes and sketches of how the engine comes apart. Page 113.

T F 11. Metric measuring tools may be used to determine wear of engine components. Page 109.

FOUR-STROKE CYCLE ENGINE MAJOR SERVICE AND OVERHAUL

BASIC COMPONENTS

Figure 42

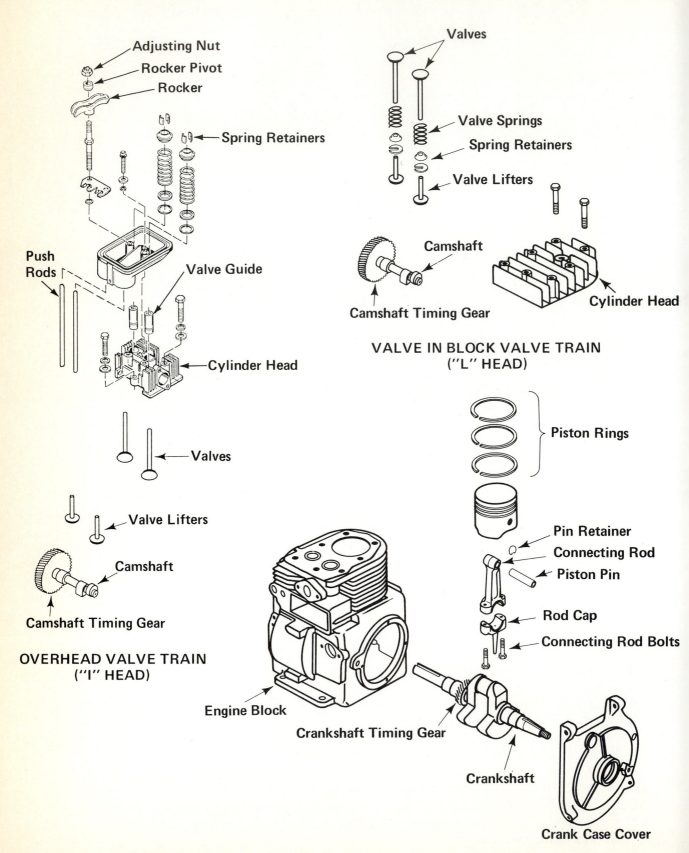

Adjusting Nut
Rocker Pivot
Rocker
Spring Retainers
Push Rods
Valve Guide
Cylinder Head
Valves
Valve Lifters
Camshaft
Camshaft Timing Gear

OVERHEAD VALVE TRAIN ("I" HEAD)

Valves
Valve Springs
Spring Retainers
Valve Lifters
Camshaft
Camshaft Timing Gear
Cylinder Head

VALVE IN BLOCK VALVE TRAIN ("L" HEAD)

Piston Rings
Pin Retainer
Connecting Rod
Piston Pin
Rod Cap
Connecting Rod Bolts
Engine Block
Crankshaft Timing Gear
Crankshaft
Crank Case Cover

BASIC COMPONENTS

ENGINE BLOCK

Most all internal-combustion engines have similar basic components. The *engine block* is the major component of all engines. This unit contains all the moving parts necessary to convert heat energy (burning gases) to mechanical energy (rotating crankshaft). All engines have a smooth, round cylinder that permits the piston to move up and down. The crankshaft turns on main bearings, which are also located in the engine block.

In most piston-type engines the piston is connected to the connecting rod by a piston pin. The other end of the connecting rod is connected to the crankshaft at the connecting rod journal.

Piston rings are fitted on the piston to prevent the loss of compression and power. In addition to sealing the cylinder, the rings also clean the cylinder wall of excess oil and exhaust deposits and transfer some of the heat from the cylinder wall to the oil.

Four-stroke engines require components that permit operation of poppet valves to control the movement of intake and exhaust gases. The camshaft is driven by a gear on the crankshaft at one-half the crankshaft speed. This gear ratio enables each camshaft lobe to move its lifter and thereby open its valve one time while the crankshaft completes two revolutions. The valve spring maintains constant tension on the valve. When the valve is opened, it is moved against the spring tension. The spring causes the valve to return to its seat. The valve will seal as long as the sealing surfaces of the valve and seat are in good condition. The two basic valve arrangements, "I" head and "L" head are shown in the accompanying drawing. The "I" head employs more valve train parts, but it is more efficient in terms of power production.

A constant supply of oil is delivered to the points of friction by a pump and splash system on some engines and entirely by a splash system on others. Seals on the crankshaft and gaskets on the bolt on components prevent leakage of the oil. Gaskets and seals should not be reused. Always install new gaskets and seals when assembling an engine.

Power of the four-stroke engine is wasted when wear occurs on the internal parts. Most wear occurs on the piston rings, piston, cylinder, and valves. Improper maintenance greatly increases wear. If the engine is run when it is low on oil, it wears very rapidly and can be ruined in a short time.

Low compression and knocking inside the engine usually are signals that an overhaul is needed. When overhauling the engine, check the clearances and either replace or recondition worn components. Piston rings should *always* be replaced if the piston is removed.

COMBUSTION CHAMBER AND HEAD SERVICE

Figure 43

<u>CAUTION</u>: SAFETY GLASSES SHOULD BE WORN WHEN REMOVING CARBON FROM ENGINE COMPONENTS.

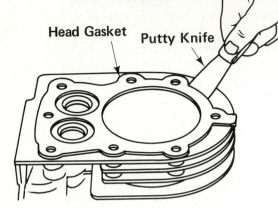

Head Gasket   Putty Knife

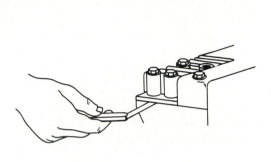

CLEANING GASKET FROM ENGINE

CHECKING FOR WARPAGE

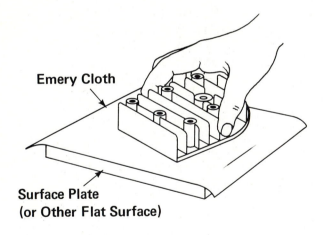

Emery Cloth

Surface Plate
(or Other Flat Surface)

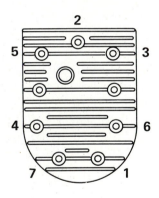

RESURFACING HEAD

EXAMPLE OF HEAD BOLT TIGHTENING SEQUENCE

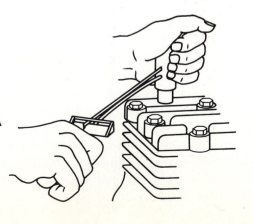

TIGHTEN CYLINDER HEAD BOLTS WITH A TORQUE WRENCH

COMBUSTION CHAMBER
AND HEAD SERVICE

1. Remove the air shroud and the parts that cover the cylinder head. Be sure to mark any wires that are disconnected. If the fuel tank is to be removed, shut off the fuel at the tank or plug the line. Clean dust from fins on the block and cylinder head. *Caution:* Wear safety glasses!

2. Remove the spark plug. (The valve springs retainers, springs, and rockers must be removed on some "I" head engines.)

3. Remove the bolts that secure the head to the engine. *Caution:* Some engines have bolts of different lengths. Note the location of the longer bolts.

4. Remove the old head gasket and clean the carbon from the head.

5. With a putty knife or carbon scraper, remove all carbon from the piston head and block surfaces. Carbon build up can reduce horsepower up to 35%.   **Check Point** _____

6. Check for warpage of the block and/or head by placing the head on the engine without using a gasket. With a feeler gauge, check the maximum space between the head and the block. If the gap is more than .005 inch (.127mm), the head should be checked on a surface plate. If the warpage is in the head, the head should be resurfaced or replaced. If the warpage is in the block, the block should be resurfaced. To resurface the head or the block, place a sheet of medium grit emery cloth over a surface plate or other flat steel surface. (The table of a power saw will work.) Move the head or block over the emery cloth until it shows a true gasket surface. The head gasket will permit a slight amount of variation between the cylinder head and the block mating surfaces.

7. Make certain that the block and head are clean and place a new head gasket on the block. *Carefully* check the gasket to be sure that it aligns properly with the bolt holes and the block surface. Place a small amount of graphite grease on the threads of the head bolts and install the head on the engine. (If graphite grease is not available, scrape a soft lead pencil over the threaded portion of the bolts.) Make certain that all bolts are installed in the proper locations. (If the bolts were accidentally mixed, check with a small nail to determine where the longer bolts are needed.)

8. Snug all head bolts by hand. Torque the bolts to one-third torque specifications following the tightening sequence shown for your bolt arrangement. Refer to the manufacturer's manual for the torque specifications and tightening sequence. If these data are not available, use a criss-cross pattern as shown in the head bolt tightening sequence. Retighten all bolts to two-thirds torque specifications. Tighten to specifications and go over them again to be sure they are properly torqued.

9. Clean, gap, and replace the spark plug. (It is best to install a new gasket under the spark plug.) Torque the plug to 20 foot pounds (27 NM). If a torque wrench is not available, thread the spark plug in until it is snug against the new plug gasket. Tighten the plug one-third turn, and it will be at the approximate torque recommendation.

10. Replace the air shroud and other components.
                              Check Point _____

ENGINE OVERHAUL
PRELIMINARY STEPS

Figure 44

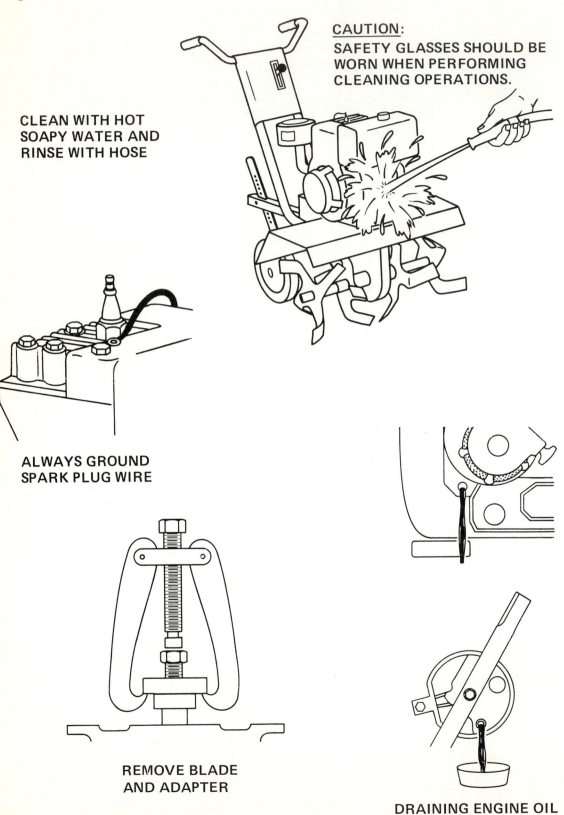

CAUTION:
SAFETY GLASSES SHOULD BE
WORN WHEN PERFORMING
CLEANING OPERATIONS.

CLEAN WITH HOT
SOAPY WATER AND
RINSE WITH HOSE

ALWAYS GROUND
SPARK PLUG WIRE

REMOVE BLADE
AND ADAPTER

DRAINING ENGINE OIL

ENGINE OVERHAUL
PRELIMINARY STEPS

The engine should be overhauled if there is poor compression because of worn cylinder components that reduce compression or cause high oil consumption. Complete disassembly also will be necessary to correct problems such as a "locked engine" or one which "knocks." Diagnosing these problems is explained in the Compression Testing section.

On most engines it is best to do a complete overhaul even though the diagnosis shows only problems with the valves.

1. *Cleanliness.* Clean the engine before starting disassembly. Clean all parts so that they can be inspected and accurately measured. Keep the work area clean and orderly.
2. *Ordered Procedure.* Keep track of how the engine comes apart— where each unit attaches and the mating of moving components. Draw sketches and make notes on the wiring connections and carburetor linkage.
3. *Working to Specifications.* All components must be checked carefully for wear and failure. During reassembly all parts must fit properly and all bolts must be tightened to the proper tension.
4. *Lubrication.* When any engine is reassembled, all internal parts must be well oiled or the engine will be damaged the instant that it is started.

The following special tools are needed when completing an overhaul:

Dial indicator
0–1″ micrometer
1–2″ micrometer
2–3″ micrometer (Metric micrometers may be substituted)
Telescoping gauges (0–1″, 1–2″, 2–3″)
Ring ridge remover
Piston ring compressor
Valve spring compressor
Valve reseating tool

RING RIDGE REMOVAL AND BASIC DISASSEMBLY

Figure 45

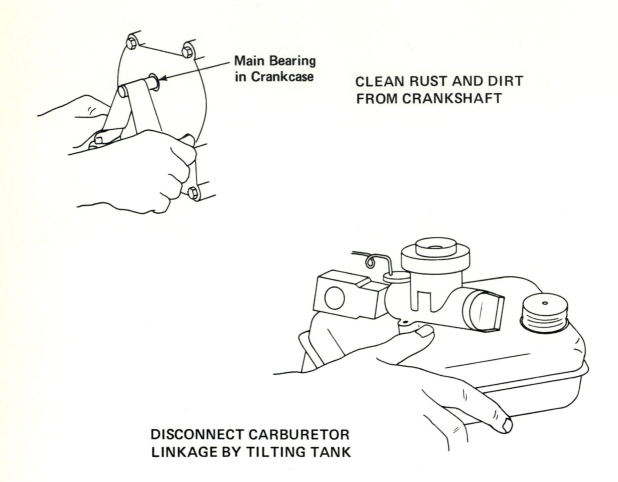

Main Bearing
in Crankcase

**CLEAN RUST AND DIRT
FROM CRANKSHAFT**

**DISCONNECT CARBURETOR
LINKAGE BY TILTING TANK**

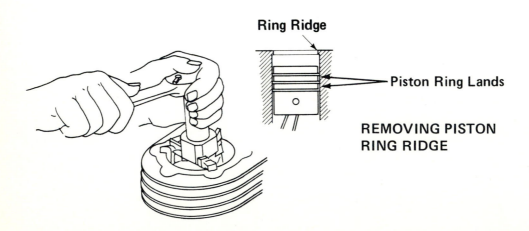

Ring Ridge

Piston Ring Lands

**REMOVING PISTON
RING RIDGE**

RING RIDGE REMOVAL AND
BASIC DISASSEMBLY

OVERHAUL PROCEDURE

1. Clean the outside of the engine with a brush. If the engine is oily, apply engine cleaner on the engine and rinse with water. Hot, soapy water can be used to clean engines that aren't caked with grease. *Caution:* Wear safety glasses when performing cleaning operations. *Do not use gasoline to clean the engine!*

2. Remove the spark plug wire and attach it to the cylinder head or block. It must be located so that a spark cannot reach the spark plug. By doing this, there is no chance that the engine will fire when either the blade or drive mechanism is being removed.

3. Drain the oil from the engine. Most engines have a plug in the base for draining the oil. Some engines can be drained by tilting the engine and letting the oil flow out the oil filler opening.

4. Remove the drive mechanism or blade.

5. Remove the engine from the tiller, mower, or tractor unit. Most engines are attached by three or four bolts at the engine base.

6. Complete Steps 1 through 6 of the Head Service section.

7. Clean all rust and dirt from the drive end of the crankshaft as shown in the accompanying illustration. On many engines the main bearing must slide over the end of the crankshaft. Any rust or roughness on the shaft will damage the main bearing when the crankcase cover is removed.

8. Remove the flywheel from the magneto end of the crankshaft. If the ignition points are operated by a crankshaft-driven plunger, remove the plunger. (See Ignition Service section for details of flywheel removal.)

9. Remove the intake manifold and carburetor assembly. Be sure that the fuel is shut off at the tank and disconnect the fuel line on models with remote tanks. *Note:* Do not bend the governor linkage! Unbolt the carburetor or manifold and rotate the assembly to disconnect the linkage.

10. Remove the muffler. Penetrating oil will help loosen rusted threads.

EXTERNAL CRANKSHAFT CHECKS AND VALVE REMOVAL

Figure 46

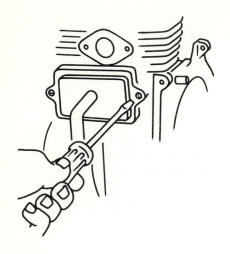

REMOVING VALVE COVER

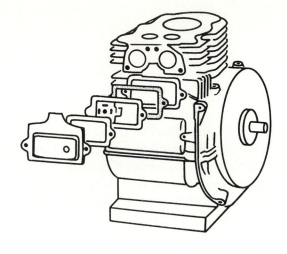

KEEP BREATHER COMPONENTS IN ORDER

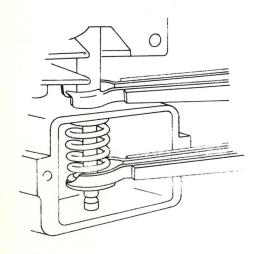

VALVE REMOVAL

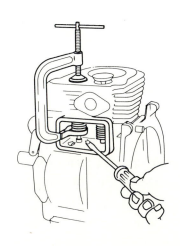

VALVE REMOVAL

EXTERNAL CRANKSHAFT CHECKS AND VALVE REMOVAL

11.   Remove the ring ridge from the top of the cylinder. If a ring ridge remover is not available, carefully remove the ridge by hand with medium grit emery cloth. Failure to remove the ridge may damage the piston when it is reassembled with new piston rings. The sharp corner on the new rings will catch on the ridge and break the piston ring land beneath the first ring groove.

Check Point _____

12.   Remove the valve cover. Some covers contain a filter unit and serve as a crankcase breather. Be careful to observe the assembly order of units that contain filter and breather valve components.

13.   Remove the valves by compressing the valve spring and removing the retainer with pliers. If a valve spring compressor is not available, two screwdrivers can be used to compress the springs of valves in block engines.

Note: On some overhead valve engines, the valve rocker cover, spring retainers, springs, and rockers are removed prior to removal of the cylinder head.

Keep the valve springs with their respective valves. On some engines the exhaust valve has a spring that is heavier than the intake valve spring.

14.   Check the end play of the crankshaft by clamping a dial indicator to the crankshaft with the pad resting against the crankcase. Move the crankshaft in and out. The indicator will show the end play in thousandths of an inch.

Figure 46 (cont.)

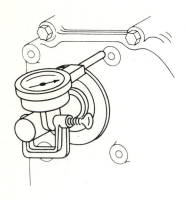

CHECKING CRANKSHAFT END PLAY WITH DIAL INDICATOR

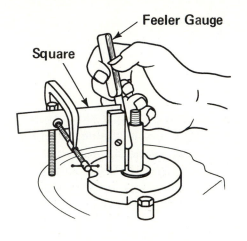

CHECKING FOR BENT CRANKSHAFT WITH SQUARE AND FEELER GAUGE

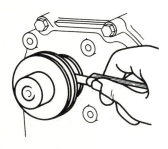

CHECKING CRANKSHAFT END PLAY WITH FEELER GAUGE

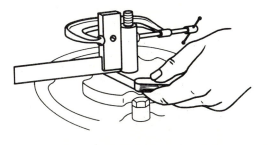

CHECKING CRANKSHAFT END PLAY WITH A SQUARE AND FEELER GAUGE

EXTERNAL CRANKSHAFT CHECKS AND VALVE REMOVAL (cont.)

If a dial indicator is not available, the end play can be checked with a feeler gauge. On some engines the feeler gauge can be placed between the drive pulley and the crankcase. Measure the gap when the crankshaft is pulled "out" and then measure the gap when it is pushed "in." The difference between the two measurements is the crankshaft end play.

Another method of checking the end play is to clamp a square on the crankshaft with a C-clamp. Take a measurement with the feeler gauge when the crankshaft is pulled "out" and another measurement when it is pushed "in." The difference between the measurements is the crankshaft end play.

Record the crankshaft end play in the Data Block below. If the end play is not correct, it can be adjusted when the engine is being reassembled.

If the engine is from a rotary lawnmower, check for a bent crankshaft. To make this check, mount a dial indicator on the engine with the pad against the crankshaft. Slowly rotate the crankshaft and observe the movement of the indicator needle.

An alternate method is to clamp a bar or square on the engine so that it is very close to the end of the crankshaft. Rotate the crankshaft and measure the difference in the gap between the bar and the crankshaft with a feeler gauge. A difference in the gap indicates a bent crankshaft.

A crankshaft with more than .005" (.127 mm) run out should be replaced.

Check Point _____

Data Block

	Actual	Specifications
Crankshaft end play	_____	_____
Crankshaft runout	_____	.005" Maximum (.127 mm)

NOTES: _____

_____

_____

_____

_____

_____

_____

_____

_____

OPENING THE CRANKCASE
AND PISTON REMOVAL

Figure 47

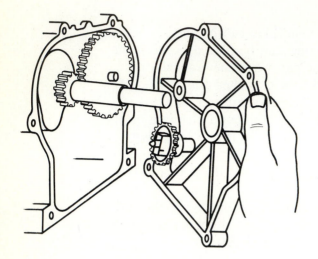

REMOVING CRANKCASE BASE

MARKS ON CRANKSHAFT
AND CAMSHAFT TIMING GEARS

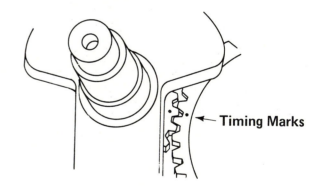

Timing Marks

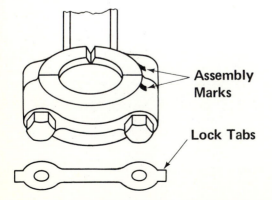

Assembly
Marks

Lock Tabs

CONNECTING ROD AND
CAP REFERENCE MARKS

OPENING THE CRANKCASE AND PISTON REMOVAL

15. Position the engine so that it is easy to remove the removable crankcase side or base.

16. Remove the bolts that attach the crankcase side or base cover. Carefully remove the cover. If a gasket sealer was used, it will be necessary to tap the cover with a hammer to "break it loose."

Don't discard the old crankcase side cover gasket because it must be measured when selecting the gasket to be used in reassembly.

On engines that have a side cover, check for timing marks on the camshaft and crankshaft gears. Some engines have a mark on the camshaft gear that aligns with a mark on the crankshaft counterweight. Other engines have a mark on the camshaft gear that is aligned with the crankshaft gear key or a mark on the crankshaft gear.

Rotate the crankshaft until the two timing marks align. If no marks are visible, wipe the oil from the matching teeth on the two gears and paint a mark on each gear.

Check Point _____

17. Remove the camshaft oil distributor or slinger if one is used.

Carefully remove the camshaft and valve lifters. Mark the lifters with masking tape so that they can be reinstalled in the same location.

Note: The above step can be omitted on an engine that has a removable base.

18. Study the connecting rod and cap to identify marks or reference tabs. If no locator marks or tabs are visible, make a small punch mark on the rod and the rod cap on the camshaft side of the engine. Note these data at the end of this chart.

With a punch and hammer or pliers, straighten the sheet metal locks on the connecting rod nuts. (Some engines use self-locking nuts instead of a sheet metal lock.)

Remove the connecting rod nuts or bolts with a socket wrench or box-end wrench. Carefully push the piston-rod assembly out the top of the cylinder.

Caution: DO NOT permit the rod bolts to "bang" against the crankshaft.

DO NOT place any tools against the bearing portion of the connecting rod.

19. Once the piston assembly is removed from the engine, reinstall the connecting rod cap on the connecting rod. Remove the crankshaft from the engine block. Make certain that the bearing surfaces of the crankshaft aren't nicked by careless handling. The main bearing retainer or side of the block must be unbolted on an engine that has a removable base.

Figure 47 (cont.)

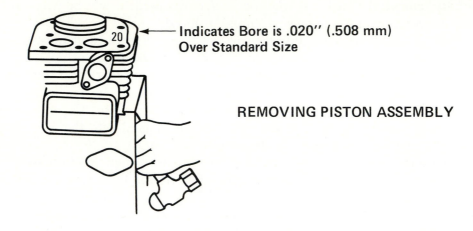

Indicates Bore is .020'' (.508 mm) Over Standard Size

REMOVING PISTON ASSEMBLY

THE MARKING ON THE PISTON INDICATES THAT THE BLOCK HAS BEEN MACHINED .020'' OVER STANDARD

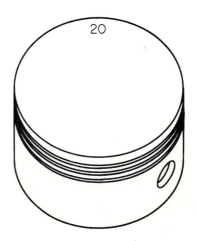

20. Clean all parts, including the inside of the block, in parts cleaning solvent or in kerosene. DO NOT use gasoline to clean parts because gasoline leaves a chalky deposit when it evaporates and it is a fire hazard. *Caution:* Wear safety glasses when performing cleaning operations.

21. Inspect the head of the piston for its size. If the engine cylinder has not been resized (bored), the top of the piston will have no marking or STD stamped upon it. STD means that the cylinder is standard sized. A cylinder that has been resized will have the amount of diameter-over-standard stamped on the top. Oversizes are usually .010″, .020″, or .030″ (.254, .508, or .762 mm) greater than the manufacturer's standard bore for the engine.

CYLINDER MEASUREMENTS AND CHECKS

Figure 48

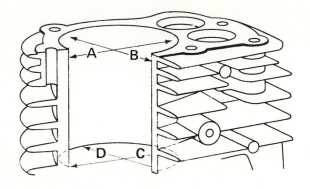

PLACE WHERE CYLINDER MUST BE MEASURED

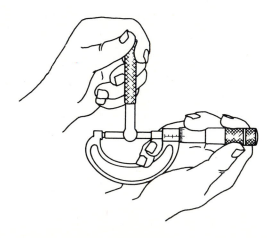

CHECKING CYLINDER WEAR WITH A TELESCOPING GAUGE AND A MICROMETER

CYLINDER MEASUREMENTS AND CHECKS

CYLINDER CHECKS

22. The cylinder must be checked for wear and scores or scratches. There are three different kinds of wear that can occur in the cylinder—oversize, out-of-round, and taper.

The three types of wear are checked by taking precision measurements at two locations in the cylinder—three-quarters of an inch from the top and three-quarters of an inch from the bottom.

The measurements can best be taken with a telescoping gauge and outside micrometer, an inside micrometer, or a cylinder dial gauge.

First, take two measurements three-quarters of an inch from the top of the cylinder and at 90° to each other (measurements A and B in the first illustration). The difference between the smaller of the two measurements and the original bore of the engine (2 inches, 2-1/4 inches, 2-5/16 inches, etc.) represents cylinder wear. Record the cylinder wear in the Data Block.

The difference between the two measurements taken three-quarters of an inch from the top of the cylinder (measurements A and B in the first illustration on the left) represents the out-of-round of the cylinder at the top. Record the out-of-round in the Data Block.

The difference between the measurements taken at right angles to the crankshaft (measurements A and C in the first illustration) is the cylinder taper. Record the cylinder taper in the Data Block.

If precision measuring instruments are not available, the cylinder taper can be measured with a new or used piston ring and a feeler gauge (third illustration). To measure taper in this manner, carefully remove a piston ring from the piston and place it in the cylinder. With the head or skirt of the piston push the ring down until it is three-quarters of an inch from the top of the block. Make sure that the ring is square with the cylinder.

Measure the gap between the ends of the ring with a feeler gauge. (If the measurement exceeds .030″ [.76 mm], the cylinder has excessive wear and may need to be resized.)

Move the ring to three-quarters of an inch from the bottom of the cylinder and measure the gap between the ends of the ring with the feeler gauge. Subtract the reading taken at the bottom from the reading taken at the top and then divide by 3. This will give the taper in thousandths of an inch.

Compare the cylinder wear, out-of-round, and taper to the manufacturer's specifications. If any wear exceeds the maximum recommended by the manufacturer, the cylinder should be resized by a machine shop.

Cylinders are usually resized to .010″, .020″, or .030″ (.254, .508, or .762 mm) over the standard bore.

If manufacturer's data are not available, the following data can be used as a guide to determine whether or not the cylinder should be resized:

Maximum oversize diameter	.003″ (.076 mm)
Maximum out-of-round	.005″ (.127 mm)
Maximum taper	.004″ (.102 mm)

CYLINDER MEASUREMENTS AND CHECKS (cont.)

Figure 48 (cont.)

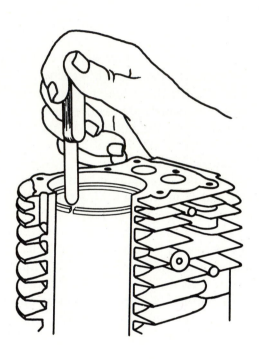

CHECKING CYLINDER WEAR WITH A PISTON
RING AND A FEELER GAUGE

CYLINDER MEASUREMENTS
AND CHECKS *(cont.)*

Inspect the cylinder for deep scores or scratches. A deep scratch will cause compression loss and oil burning even though the cylinder wear is not excessive.

A badly scored or scratched cylinder should be resized.

Resizing is accomplished by increasing the cylinder size with special cylinder hones, a boring bar, or a metal cutting lathe. Resizing procedures recommended by the resizing equipment manufacturer should be followed carefully.

Data Block

Cylinder wear _____

Cylinder out-of-round _____

Cylinder taper _____

Check Point _____

PISTON CHECKS

Figure 49

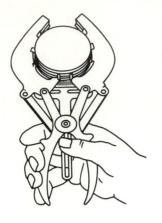

REMOVING PISTON RINGS

<u>CAUTION</u>:
SAFETY GLASSES SHOULD BE WORN WHEN CLEANING PISTONS!

CLEANING RING GROOVES
WITH RING GROOVE CLEANER

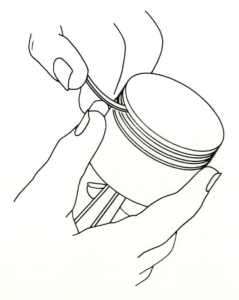

CLEANING RING GROOVE
WITH BROKEN RING

PISTON CHECKS

23. If the cylinder is resized (rebored), a new piston must be installed.

If the engine is not to be rebored and the piston is to be reused, the old rings must be removed. To remove the rings, use a ring expander as shown in the accompanying figure. If a ring expander is not available, carefully expand the ring by hand and move it up over the top of the piston. AVOID SCRATCHING the ring lands with the ends of the ring. Remove the other rings in the same manner.

Clean the carbon from the top of the piston with a carbon scraper or putty knife. DO NOT use a wire brush to clean the piston because it will wear away part of the ring lands. Stuck piston rings, broken rings, severe piston burning, and top groove wear can be the result of improper combustion or detonation. Careful service of the ignition system and the use of proper fuel will prevent such problems from occurring.

A ring groove cleaning tool is used to remove carbon from the bottom of each ring groove. Adjust the cutter head so that the proper sized cutter will be in line with the ring groove when the tool is placed on the piston. Place the cutter in the groove to be cleaned and adjust the tool so that spring tension keeps the tool under tension. Rotate the tool by hand until all the carbon is removed from the groove. This procedure is followed for each ring groove.

If a ring groove cleaning tool is not available, a broken piston ring can be used to scrape the carbon from the ring grooves. If a broken ring is to be used, the scraping end should be square so that it scrapes evenly along the bottom of the groove. If the carbon is not cleaned from the grooves, it can cause the new rings to bind in the cylinder when the engine is reassembled.

Wash the piston in solvent and carefully inspect it for burned areas on the piston lands or deep scores, scratches, or cracks on the skirt.

If the piston has any of the above defects, it should be replaced. Scoring or scuffing usually results from inadequate lubrication or a dirty cooling system. If scores are present, carefully examine the cooling fins and passages on the block. Dirt-clogged passages can cause the engine to overheat. Poor lubrication can also cause cylinder scoring. Inspect the lubrication system for broken or damaged components.

Check the ring groove for wear by placing a new ring in the groove and checking the remaining space with a feeler gauge. If a .007″ (.178 mm) or larger feeler gauge can be inserted between the ring and the piston, the piston should be replaced. Excessive wear in the ring groove can cause the rings to "pump" oil to the combustion chamber. Such wear can also cause the piston rings to break.

Check Point _____

Figure 49 (cont.)

SCORED PISTON

BURNED PISTON

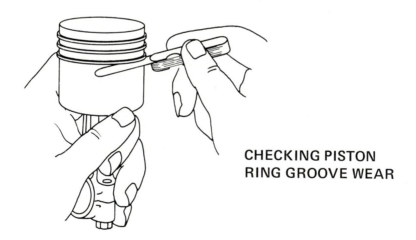

CHECKING PISTON
RING GROOVE WEAR

SCRATCHED RING

24. Inspect the old piston rings. Careful inspection may indicate the causes of wear or damage.

Fine vertical scratches on the ring faces show that dirt has entered the engine along with the intake air. Failure to correct the problem will cause the new rings to wear out quickly. The usual source of such dirt is the air cleaner. The air cleaner must be serviced periodically. (See Air Cleaner Service section, page 105.) It must be connected tightly to the carburetor to prevent entry of dirt.

Deep scratches or scoring on the rings are normally the result of an engine that has overheated. Overheating can be caused by insufficient cylinder lubrication, clogged cooling fins, or air passages on the block, incorrect combustion, or insufficient ring or piston clearance.

PISTON PIN AND
CONNECTING ROD CHECKS

Figure 50

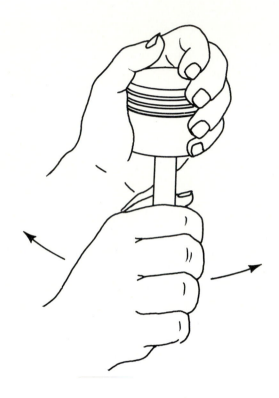

CHECKING PISTON PIN WEAR

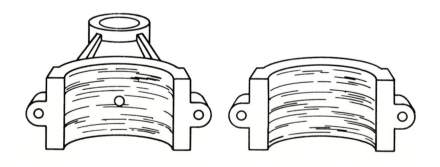

SCORED ROD BEARING

PISTON PIN AND CONNECTING ROD CHECKS

PISTON PIN CHECKS

25. Piston pin wear can be checked by holding the piston firmly in one hand while attempting to rock the rod with the other hand. The rod must be rocked in line with the piston pin. It is normal for the rod to slide freely across the piston pin. Don't confuse this with piston pin wear.

If looseness is detected, the piston pin should be removed. Before removing the pin, note identifying marks or mark the piston and rod with a punch mark so that the two can be reassembled properly. Remove the piston pin retainer with small nose pliers and remove the piston pin. Measure the piston pin with a micrometer. Compare the measurement with the standard size recommended by the manufacturer. If the pin is worn over .0005″ (.0127 mm) out-of-round or below the minimum specified by the manufacturer, the pin should be replaced.

Check Point _____

Check the pin bore in the piston for wear with a telescoping gauge and a micrometer. If the bore is over .0005″ (.0127 mm) out-of-round or oversize, the bore should be machined for an oversize pin or the piston should be replaced.

Oversize piston pins are available for some engines. To install an oversize pin, the piston and connecting rod bore must be reamed to fit the new pin. This machining operation should be performed by a machine shop equipped to complete the precision reaming required.

Some manufacturers include new piston pins with new pistons. The fit between the rod and the new pin should be checked. If the rod is loose on the new pin, the rod should be replaced.

CONNECTING ROD CHECKS

26. Inspect the connecting rod for wear. If there was looseness when the piston-rod assembly was checked for wear, the pin bore in the rod should be measured with a telescoping gauge and micrometer. If the pin bore in the rod is over .0007″ (.0178 mm) out-of-round or is scored, the rod should be replaced.

The crankpin bearing in the connecting rod should be inspected carefully for pits and scratches. If the bearing surface is pitted or scored, the connecting rod should be replaced. If there are wear spots on diagonally opposite points of the rod and the rod cap, the connecting rod is bent and should be replaced. It is NOT recommended that the connecting rod cap be filed to compensate for wear in the rod bearing or crankpin journal.

If the connecting rod was "loose" on the crankshaft when the engine was disassembled, the crankpin bearing in the connecting rod should be measured carefully with a telescoping gauge and micrometer. The bore should not be over .0005″ (.0127 mm) out-of-round. If the bore is over .0005″ (.0127 mm) over the manufacturer's standard size, the rod should be replaced.

Check Point _____

141

Figure 51

Drive End
Journal

Connecting Rod Journal

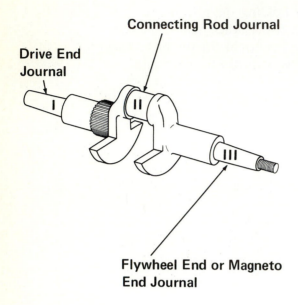

I

II

III

Flywheel End or Magneto
End Journal

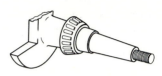

ANTIFRICTION-TYPE BEARING

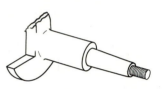

PLAIN-TYPE BEARING

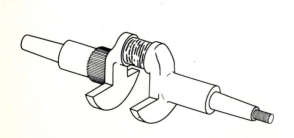

SCORED CONNECTING ROD JOURNAL

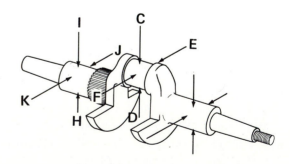

**POINTS OF MEASUREMENT FOR
CHECKING JOURNAL OUT-OF-ROUND**

CRANKSHAFT CHECKS

27. The crankshaft of the single cylinder engine incorporates three bearing surfaces. Each surface is referred to as a *journal*. On some engines the surface of the journal serves as the bearing surface. This bearing is referred to as a *plain bearing*. On some engines an antifriction bearing (taper roller or ball type) is pressed on to the crankshaft.

If the engine has plain-type bearings on the "mains" and the rod, each bearing surface must be checked carefully. Each surface should be inspected for roughness and scoring. If any of the journals is very rough, the crankshaft should be replaced. In a few cases the shaft can be polished with emery cloth and reused.

Each bearing journal should be measured carefully with a micrometer. The measurements should be taken as illustrated in the accompanying illustration.

The drive-end main journal should be checked for out-of-round by comparing the measurement H–I to measurement J–K. The difference between the measurement is out-of-round. Record the out-of-round measurement in the Data Block. If the manufacturer's data are not available, the out-of-round normally should not exceed .001″ (.025 mm).

The flywheel or magneto main bearing journal should be checked for out-of-roundness in the same manner. Record the out-of-round measurements in the Data Block. Check the connecting rod journal for out-of-round by comparing measurement C–D to measurement E–F. Record the out-of-round in the Data Block.

The taper of the bearing journals can also be checked. The taper on any plain bearing journal is measured by taking measurements at each end of the journal as shown in the accompanying illustration. The taper on journals of small engine crankshafts normally should not exceed .001″ (.025 mm). If the journal is not scored and is not out-of-round, it seldom will have "taper."

Crankshafts that exceed the wear limits or ones that are badly scored should be replaced. Some manufacturers have replacement main bearings and rods for use on undersized crankshafts. If these are available, the crankshaft can be reconditioned by a machine shop. Be certain that undersized bearings are available before having the crankshaft reground.

On engines with antifriction (taper roller or ball-type) main bearings, the bearings should be carefully inspected for nicks on the rollers or balls and races. Defective bearings can also be identified by rotating the bearing by hand. Roughness in the bearing can be felt as the bearing is turned slowly. *Note:* The bearing must be clean when it is checked for roughness.

If roughness or wear is detected, the bearing should be replaced. Some of these bearings are a press fit on the crankshaft. To remove these, an arbor press or special bearing puller is required.

Replacement procedures for this type of bearing vary. The manufacturer's recommendations should be followed when installing the new bearing. *Note:* When installing these bearings NEVER use a torch to heat the bearing.

Figure 51 (cont.)

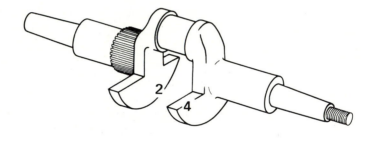

CRANKSHAFT CHECKS (cont.)

As explained earlier in the text, most manufacturers recommend that a bent crankshaft be replaced. Straightening a bent crankshaft weakens the shaft. Such a shaft could break later while in operation.

Inspect the timing gear on the crankshaft for wear and damaged teeth. Normally these gears will last the life of the engine. If there is much wear or if there are damaged gear teeth, the gear should be replaced. On a smaller engine the timing gear is made as a part of the crankshaft. It will be necessary to replace the crankshaft if the timing gear is faulty.

Data Block

	Actual	*Specifications*
Drive-end main journal out-of-round	_____	_____
Magneto-end main journal out-of-round	_____	_____
Connecting rod journal out-of-round	_____	_____
Drive-end main journal taper	_____	_____
Magneto-end main journal taper	_____	_____
Connecting rod journal taper	_____	_____

NOTES: _____

_____

_____

_____

_____

_____

_____

Check Point _____

MAIN BEARING AND
CAMSHAFT CHECKS

Figure 52

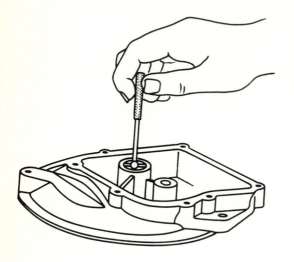

MEASURING MAIN BEARING BORE

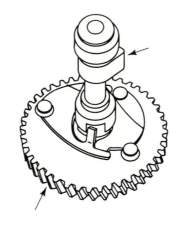

COMPRESSION
RELEASE
MECHANISM

CHECK FOR WEAR

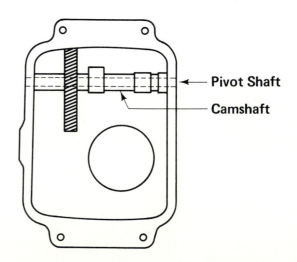

← Pivot Shaft

Camshaft

ENGINE WITH REMOVABLE BASE AND
CAMSHAFT HELD IN THE BLOCK BY A
PIVOT SHAFT

MAIN BEARING AND CAMSHAFT CHECKS

MAIN BEARING CHECKS

28. On engines that have plain-type main bearings it is important to check the bearing bores for wear.

Carefully check each bearing (one on the block and one in the crankcase cover) for scoring. If there is much scoring or pitting, the bearing should be replaced. Wear of the main bearing can be checked further with a telescoping gauge and micrometer. On some engines the bearing bore can be reamed with a reamer and a replacement sleeve can be installed. Such sleeves are *not* available from some manufacturers. If the bearing is not serviceable and a replacement sleeve is not available, the bearing can be reamed out and a sleeve can be custom-made by a machine shop. In such cases the costs of the repair must be weighed against the costs of replacing the engine or short block.

A short block includes the block, crankshaft, rod and piston assembly, valve train, and side cover. Installing a short block involves relatively little work and provides a literally new engine.

CAMSHAFT CHECKS

29. The camshaft opens the valves by operating valve lifters that push the valves off their seats. The camshaft will normally last the life of the engine if the engine oil is properly maintained.

On some engines the camshaft is easily removed once the crankcase side or base is removed. On other engines the camshaft is on a pivot shaft that runs completely across the block. To remove this camshaft requires that the pivot shaft be forced out of the block. It is not normally necessary that the camshaft be removed as a part of the overhaul.

Check the camshaft for wear by carefully inspecting each lobe for a scored or scratched surface. If there is no scoring, normally the camshaft is satisfactory. The lobes can be checked further by comparing the lobe measurements with the manufacturer's specifications.

The camshaft gear or assembly should be replaced if the gear teeth are worn, broken, or chipped. Camshaft bearing wear can be checked by comparing the shaft measurements with the standards recommended by the manufacturer.

On engines in which the camshaft is held in the block on a pivot shaft, bearing wear can be checked by feeling for looseness. To make this check, grasp the camshaft and attempt to move it toward the valve area of the engine. Then try to move it in the opposite direction. If there is much movement of the camshaft on its bearings, the camshaft should be removed and further checks made with a micrometer.

On engines that have governor units fitted to the camshaft, the governor should be checked for wear. Refer to the manufacturer's manual for data on a particular governor. Some manufacturers mount a centrifugal mechanism on the camshaft to provide compression release. On these units, check for binding of the weights and free operation of the springs.

VALVE LIFTER AND
VALVE GUIDE CHECKS

Figure 53

WORN LIFTER FACE

CLEANING VALVE GUIDE

CHECKING VALVE GUIDE
WEAR WITH DIAL INDICATOR

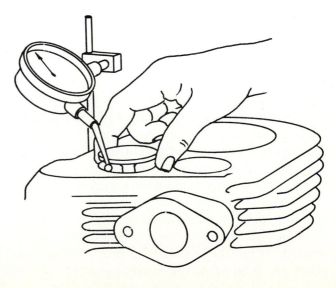

VALVE LIFTER AND VALVE GUIDE CHECKS

30. Examine the valve lifters for wear or scoring on the lifter face, which rides on the cam lobes. If the face is worn or scored, the lifter should be replaced. The face should not be remachined or ground because this will remove the hardened surface of the lifter. Without the hardened face surface, the lifter will wear very rapidly.

Before valve guides can be checked for wear, clean carbon from the guide with a valve guide brush. Check for valve guide wear by placing a new valve (a used valve will work providing the stem is not worn over .001", or .254 mm) in the guide and measuring the side play with a dial indicator. The maximum allowable clearances are listed below:

Valve Head Diameter		Maximum Side Play
Up to 1-1/4"	Intake	.005" (.127 mm)
	Exhaust	.007" (.178 mm)
Over 1-1/4"	Intake	.006" (.152 mm)
	Exhaust	.008" (.203 mm)

Another means of checking valve guide wear is to measure the valve guide with a small-hole gauge and micrometer. Compare the guide diameter one-quarter inch from the top of the guide to the specifications listed by the manufacturer.

If valve guide wear exceeds the manufacturer's recommendations, the guide should be replaced or reconditioned.

The guide can be resized by a knurling process that can be done by most automotive machine shops.

Some manufacturers have valves with oversized stems that can be installed in guides that have been reamed to the proper oversize.

Replacement guides are available for many small engines. Follow the manufacturer's recommendations when replacing the valve guide.

On overhead ("I" head) valve engines, the rocker arm, rocker arm pivot, push rods, and push rod guide must also be checked for wear. It is good practice to replace all these components if one part exhibits wear as these components "wear in" to each other.

Check Point _____

VALVE SEAT SERVICE

Figure 54

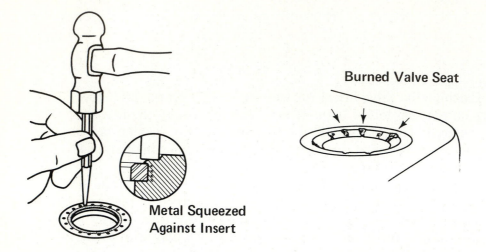

Burned Valve Seat

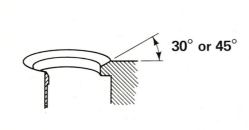

Metal Squeezed
Against Insert

30° or 45°

VALVE SEAT ANGLE

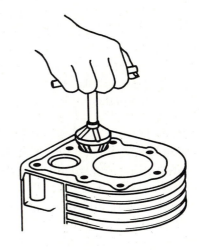

RECONDITIONING THE VALVE SEAT

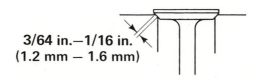

3/64 in.—1/16 in.
(1.2 mm — 1.6 mm)

VALVE PROPERLY SEATED IN BLOCK

VALVE SEAT SERVICE

31. Clean the carbon from the valve ports and valve seat area. *Caution:* Wear safety glasses when performing cleaning and valve service operations.

To ensure a good seal between the valve and the valve seat requires that the valve seat be in good condition. If the valve seat is loose in the block, it can be tightened by staking the block adjacent to the seat with a punch and hammer.

Badly burned valve seats should be replaced. Follow the recommendations of the engine manufacturer when replacing valve seats.

Valve seats that are not badly burned can be reconditioned by machining or grinding a new sealing surface. One type of seat reconditioner uses a hand-operated cutter. To use this tool, first place the proper-sized pilot in the *clean* valve guide. The proper-angled cutter (usually 45° or 30°) is then placed over the pilot. The cutter is turned in a clockwise rotation by the special T-wrench that comes with the reconditioning tool set.

Pressure is exerted on the tool as the cutter is turned. The cutting action is continued until a clean seating surface appears completely around the seat. *Caution:* The seat should be machined as little as possible. The width of the finished seating area should be from 3/64 (1.2 mm) inch to 1/16 inch (1.6 mm).

If the seating area exceeds 1/16 inch (1.6 mm) in width, it should be narrowed by removing stock from the top of the machined area. The narrowing operation is performed with a cutter that is of a lesser angle than the valve seat. (A 30° cutter can be used to narrow 45° seats.)

If much narrowing of the seat is required, be sure to check the height of the valve head when it is seated. The lower edge of the valve margin should be above the top of the valve seat (see accompanying illustration). If the valve rides too low in the seat, check to be certain that the valve margin is sufficient. If the margin is satisfactory, the problem probably is caused by the seat being too large. Before condemning the valve seat, however, place a new valve in the engine to see if it will correct the problem. If it rides properly in the seat, there is no need to install a new seat. If the new valve rides too low, a new valve seat should be installed. Refer to the manufacturer's data for valve seat replacement on a particular engine.

Figure 55

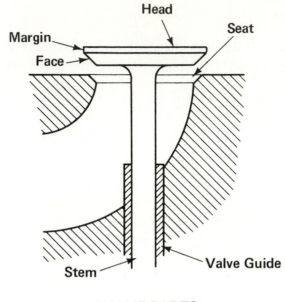

VALVE PARTS

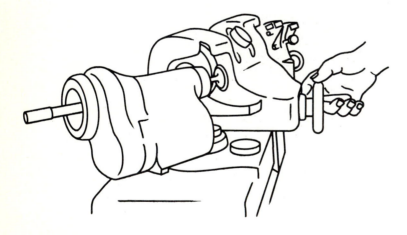

REFACING VALVE

HAND LAPPING A VALVE

VALVE SERVICE

32. Clean the intake and exhaust valves with a wire brush. Examine each valve for burned faces. If the faces are pitted or badly grooved, the valve should be replaced. *Caution:* Wear safety glasses when performing cleaning operations.

Measure the margin on each valve. On most engines, if the margin is less than 1/32″ (.794 mm), the valve should be replaced.

Check the valve stems for scoring and wear. Check for wear by measuring the valve (with a micrometer) one inch above the retainer groove or hole and one-quarter of an inch below the upper point of contact with the valve guide. If these measurements differ by .001″ (.025 mm) or more, the valve should be replaced.

If the valves are to be reused, the faces should be reconditioned by grinding on a valve refacer. Follow the recommended procedures for the valve refacer in the refacing operation. Be certain that the machine is set for the proper face angle before starting the refacing operation. Some engines use different angles for the intake and exhaust valves. If a valve refacer is not available, the operation can be performed by most automotive machine shops.

An alternate method of reconditioning the valve and the valve seat is known as *lapping.* If the valve seat and valve face are in relatively good condition, the sealing surfaces can be reconditioned by the lapping process. To do this, first place the valve in its guide in the engine. Put a small amount of lapping compound between the valve face and the valve seat. (Be sure that none of the lapping compound gets on the valve stem!) Place the suction cap of the lapping tool over the head of the valve. Rotate the valve back and forth against the seat. Frequently raise the valve off the seat to allow the compound to flow to the point of friction. Continue the lapping operation until a good sealing surface is obtained. Clean all the lapping compound from the valve and valve seat.

The final process on the valve (adjusting the clearance) can only be completed later when reassembling the engine. This operation will be explained as a part of the Reassembly Procedure.

REASSEMBLY PROCEDURE

Figure 56

CAUTION: SAFETY GLASSES SHOULD BE WORN WHEN PERFORMING THE ENGINE REASSEMBLY OPERATIONS!

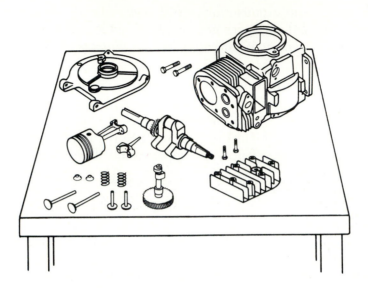

ALL COMPONENTS OF THE ENGINE SHOULD BE CLEANED AND PLACED ON A CLEAN WORK SURFACE BEFORE BEGINNING THE REASSEMBLY

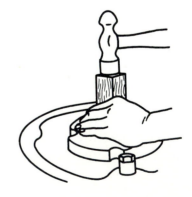

REMOVING CRANKSHAFT SEAL   INSTALLING CRANKSHAFT SEAL

REASSEMBLY PROCEDURE

When reassembling the engine, it is very important that everything—parts, work area, your hands, and tools—be clean. Surgical cleanliness is a must if the engine overhaul is to be successful. *Caution:* Wear safety glasses when performing cleaning and reassembly operations.

   1.   Rinse all engine parts in solvent and dry them with compressed air. As each part is cleaned, place it on the clean surface of the workbench. Avoid drying parts with rags because lint will collect in the small crevices. If your work on the engine is interrupted for any length of time, cover the engine components with a clean, lint-free cloth.

   2.   Install new crankshaft seals in the engine block and the cover housing. The old seals can be pried out with a large screwdriver or they can be driven out with a punch and hammer.

   When new seals are being installed, be certain that the sharp edge or lip of the seal is toward the inside of the engine. Some manufacturers recommend that a film of sealer (Permatex #2) be applied to the outer part of the seal (between the seal and the casting) to prevent oil leakage between the seal and the housing. Seat the seal in the housing with a seal driver (or block of wood) and a hammer.

   3.   If the piston, piston pin, or connecting rod is being replaced, proceed with installation of the piston on the connecting rod. First, match the old components that you marked prior to disassembly. Position the new replacement components adjacent to the old parts. Assemble the piston and rod so that it is just like the old parts were prior to disassembly.

Check Point _____

   4.   If the camshaft was removed and was the kind held in the engine block on a pivot shaft, proceed with its replacement. First, clean the block lifter holes and camshaft bearings. Lubricate these surfaces with motor oil (SAE 20 or 30, SE or SF classification). Install the lifters in their respective bores and place the camshaft in the block. Install the camshaft pivot shaft and expansion plugs in the block.

REASSEMBLY PROCEDURE
(cont.)

Figure 56 (cont.)

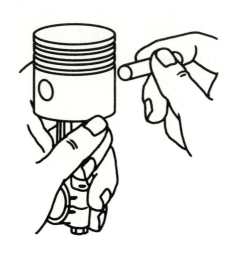

INSTALLING PISTON ON ROD

REASSEMBLY PROCEDURE
(cont.)

5. Wipe the main bearing surfaces in the engine block and lubricate them with motor oil.

Wipe the main bearing journals on the crankshaft, apply film of oil, and install the crankshaft in the engine. *Carefully* guide the crankshaft through the seal on the magneto or flywheel side of the engine. *Note:* The following procedure is only for an engine whose camshaft is held in the engine on a pivot shaft. If your engine has a removable side crankcase cover or base containing a main bearing, disregard the remainder of this step. Align the timing marks on the crankshaft and the camshaft gears. Refer to the Data Block in the disassembly procedure for the crankshaft end play recorded prior to disassembly of the engine. Compare this end play measurement to the manufacturer's data. If the end play is within specifications, select a new bearing retainer cover gasket of the same thickness as the one that was removed from the engine during disassembly. Install the gasket and bearing retainer. If the end play of the crankshaft was more than recommended by the manufacturer, select a thinner gasket and install the retainer.

Using the dial indicator, check the crankshaft end play with the new gasket in place. If the end play is not within specifications, remove the retainer and select a gasket that will provide the proper end play. If a dial indicator is not available, check the end play with a feeler gauge as shown in the disassembly procedure. If manufacturer's data are not available, adjust the end play between .004″ (.102 mm) and .012″ (.305 mm).

INSTALLING PISTON RINGS

Figure 57

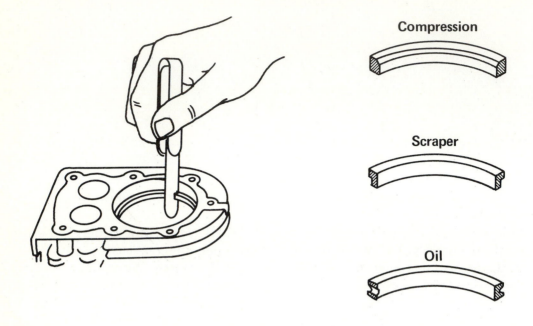

CHECKING END GAP OF NEW PISTON RING

EXAMPLES OF PISTON RINGS

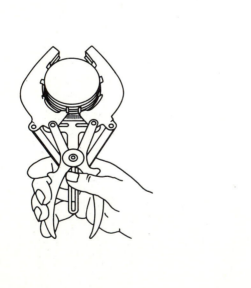

INSTALLING RINGS ON PISTON

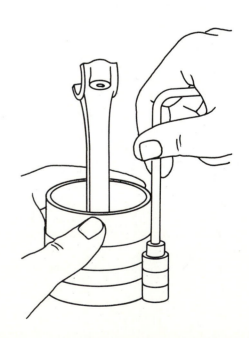

INSTALLING PISTON RING COMPRESSOR

INSTALLING PISTON RINGS

6. Place one of the new piston rings in the cylinder and push it down to approximately one-half inch from the bottom of the cylinder with the piston. Check the gap between the ends of the ring with a feeler gauge. Compare this end gap clearance with the minimum ring end gap clearance recommended by the manufacturer. The end gap must *not* be less than the minimum clearance specified or the rings will tend to seize in the cylinder. (If the manufacturer's data are not available, allow .003″ (.076 mm) end gap clearance for each inch of cylinder diameter.)

If the end gap clearance is less than recommended, carefully file the end of the ring to obtain the necessary clearance. If it is necessary to file the end of one ring, it will also probably be necessary to file the other rings.

Check Point _____

7. Install the new piston rings on the piston. *Caution:* Safety glasses should be worn when installing rings on the piston. *Always* install new piston rings in an engine that has been disassembled. Used rings cannot seat in properly and will cause oil consumption and compression loss.

When installing the piston rings, carefully follow the instructions included with the new piston rings. Lubricate the ring grooves on the piston before installing the rings. Use a piston ring expander when installing the new rings. The rings can be "stretched" if they are installed by hand-springing them over the piston.

The widest ring is the *oil ring,* which fits in the bottom groove of the piston. Some manufacturers use a thin metal expander behind this ring. The middle ring is referred to as the *scraper ring* or the *center compression ring.* It normally has a groove on its outside edge. This groove must be down when the ring is assembled on the piston. The top ring or *compression ring* normally has a bevel on its inside edge. This bevel must be up when the ring is installed on the piston.

Check Point _____

8. Lubricate the piston and rings thoroughly with motor oil. Rotate the piston rings so that the end gaps of the rings are not in line. Install the piston ring compressor over the piston and tighten. Continue to tighten the compressor until the bottom edge of the compressor is against the piston.

INSTALLING
THE PISTON ASSEMBLY

Figure 58

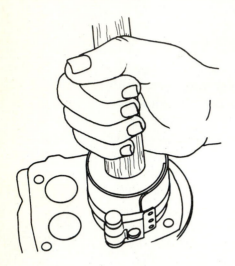

INSTALLING PISTON ASSEMBLY

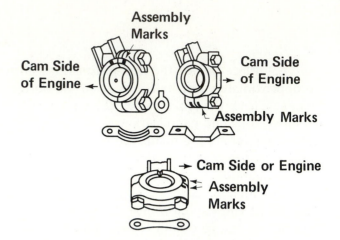

Assembly Marks

Cam Side of Engine ←

Cam Side of Engine →

Assembly Marks

→ Cam Side or Engine

Assembly Marks

EXAMPLES OF MATCH MARKS AND ROD BOLT LOCKS

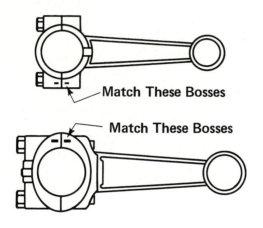

Match These Bosses

Match These Bosses

TIGHTEN ROD BOLTS TO SPECIFIED TORQUE

CHECK CONNECTING ROD BEARING CLEARANCE WITH PLASTIGAUGE

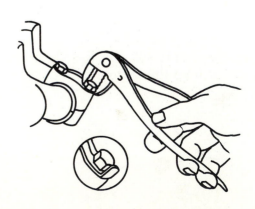

LOCKING ROD BOLTS

INSTALLING
THE PISTON ASSEMBLY

9. Wipe the cylinder with a clean cloth. Lubricate the cylinder with motor oil and rotate the crankshaft to place the rod journal opposite the cylinder. Place the piston assembly in the bore. *Make certain* that the marks on the connecting rod are positioned properly in the engine.

Position a hammer handle against the head of the piston. With one hand exert pressure on the handle while bumping the end of the hammer with the other hand. Carefully guide the connecting rod so that it does not catch on the crankshaft.

10. If desired, the clearance between the connecting rod and the crankshaft can be checked at this point.

Wipe the crankshaft and connecting rod bearing surfaces. Position the rod over the crankshaft journal. Lay a piece of Plastigauge on the crankshaft and install the connecting rod cap. Carefully tighten the rod bolts to the specified torque with a torque wrench. DO NOT TURN THE CRANKSHAFT. Remove the rod cap and compare the width of the flattened Plastigauge to the scale on the Plastigauge package. The clearance should be within the specifications recommended by the manufacturer.

If the clearance is less than the minimum, check to make certain that the rod cap was installed properly on the rod and that there is no dirt or foreign material between the connecting rod and the crankshaft. If both items check out, the connecting rod is probably the incorrect rod for the engine.

If the clearance exceeds the maximum, check for dirt or foreign material between the rod and the cap at the point of connection. If no foreign material is present, remeasure the crankpin with a micrometer and compare it with the manufacturer's data. If the crankpin size is correct, either the rod is worn or else it is the incorrect rod for the engine.

If the manufacturer's rod clearance data are not available, the following specifications can be used as a guide:

.001 ″ (.025 mm) minimum clearance between rod and crankpin

.0035 ″ (.889 mm) maximum clearance between rod and crankpin

11. Thoroughly lubricate the crankpin, rod bearing, and cap bearing. Install the rod cap (oil dipper, if used), lock plate, and rod bolts or nuts. Tighten the rod bolts evenly to the manufacturer's torque specifications and bend the lock plate tabs to secure the rod bolts or nuts.

Rotate the crankshaft to ensure that there will be no binding of the rod bearing.

CAMSHAFT AND CRANKCASE COVER INSTALLATION

Figure 59

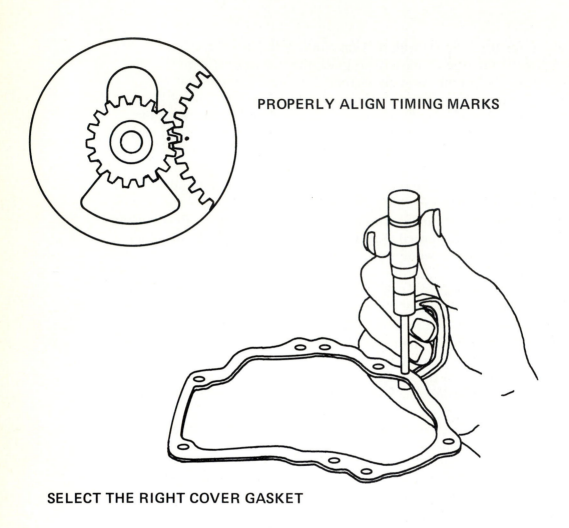

PROPERLY ALIGN TIMING MARKS

SELECT THE RIGHT COVER GASKET

WRAP CRANKSHAFT WITH
WAX PAPER TO PREVENT
DAMAGE TO THE SEAL

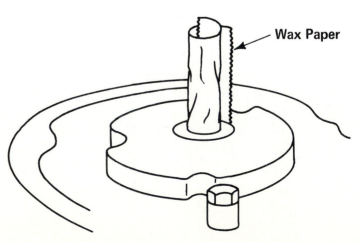

Wax Paper

CAMSHAFT AND CRANKCASE COVER INSTALLATION

12. On an engine that does not have the camshaft held in the block on a pivot shaft, lubricate the valve lifter bores and install the lifters. Lubricate the camshaft bearings and install the camshaft. MAKE CERTAIN that the timing mark on the camshaft gear and the crankshaft are matched properly.

13. Refer to the Data Block (Page 127) in the disassembly procedure for the crankshaft end play recorded prior to disassembly of the engine. Compare this end play measurement to the manufacturer's data. If the end play is within specifications, select a new crankcase cover gasket of the same thickness as the one that was removed from the engine during disassembly.

If the end play of the crankshaft was more than that recommended by the manufacturer, select a thinner gasket.

For an engine with extreme wear, some manufacturers supply a thin washer that is placed between the crankshaft timing gear and the side cover.

14. If the engine has the governor built in to the crankcase, position the governor components so that the crankcase cover can be installed.

15. On an engine that has an oil pump, check the pump for wear. Replace worn components and install the pump in the engine. With a pump oil can, apply a liberal amount of oil to the piston pin, crankshaft bearings, and camshaft.

16. Lubricate the camshaft and crankshaft bearings in the crankcase cover.

17. Install the oil slinger or oil pump.

Check Point _____

18. Carefully install the cover gaskets and cover. Be sure to work the crankshaft seal into position as the cover is installed. *Note:* It is not necessary to use a sealer on the side cover gasket. To prevent damage to the crankshaft seal use a special installing cone or wrap wax paper around the crankshaft to guide the seal over the bearing shoulder.

19. Tighten the cover retaining bolts to the specified torque.

20. Check the crankshaft end play with a dial indicator. Compare the end play of the engine to the manufacturer's data. If the end play is not sufficient, remove the cover and install a thicker gasket.

If the end play is excessive, remove the cover and install a thinner gasket.

Always use at least one gasket between the cover and the engine even though the end play slightly exceeds the maximum recommended.

If the manufacturer's data are not available, adjust the crankshaft end play between .004" (.102 mm) and .012" (.305 mm).

VALVE ADJUSTMENT
AND INSTALLATION

Figure 60

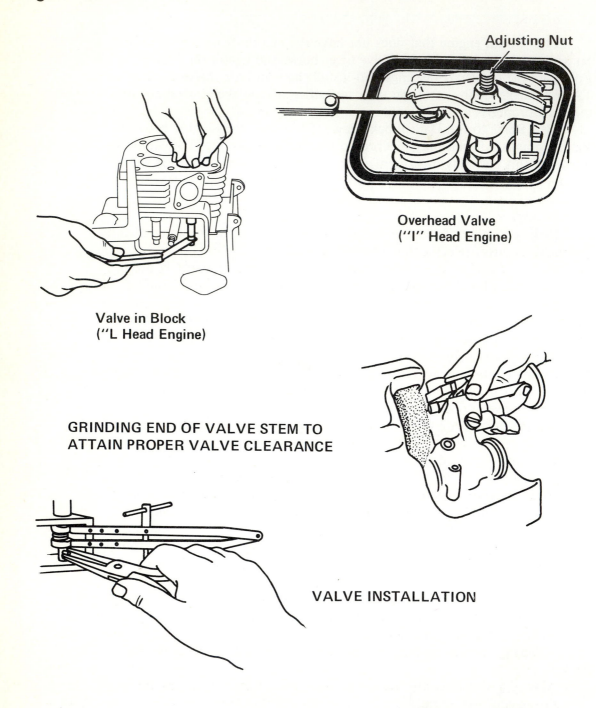

Adjusting Nut

**Overhead Valve
("I" Head Engine)**

**Valve in Block
("L Head Engine)**

**GRINDING END OF VALVE STEM TO
ATTAIN PROPER VALVE CLEARANCE**

VALVE INSTALLATION

IF VALVE SPRING COMPRESSOR IS NOT
AVAILABLE, COMPRESS SPRING IN A BENCH
VISE AND SECURE WITH WIRES AS SHOWN.
REMOVE WIRES AFTER VALVE IS INSTALLED.

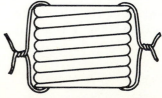

VALVE ADJUSTMENT
AND INSTALLATION

21. On an engine that has a removable crankcase base, place a new gasket on the engine block and replace the base. Tighten the bolts that secure the base. BE CERTAIN to tighten the oil drain plug.

22. For valve in block ("L" head) engines, place the intake valve in its valve guide and turn the crankshaft until the valve opens. TURN the crankshaft an additional turn. This should position the intake valve in the closed position. Hold the valve tightly closed with your thumb and check the clearance between the end of the valve stem and the valve lifter with a feeler gauge.

Check the manufacturer's specifications for the proper valve operating clearance for your engine and compare the clearance to the measured clearance of the engine. Because the valve seating surfaces have been machined, it will probably be necessary to remove some material from the end of the valve stem to achieve the right clearance.

The valve stem can be "shortened" by grinding material from the stem on a valve reconditioning machine or on a bench grinder with a V block. Grind a small amount; then recheck the clearance. If too much material is ground off, the valve will have to be replaced.

Once the intake valve operating clearance is properly adjusted, place the exhaust valve in the engine. Rotate the crankshaft until the exhaust valve opens; then rotate the crankshaft an additional turn. Check the clearance specification of the exhaust valve and adjust it in the same manner as the intake valve.

Check Point _____

Note: On overhead valve ("I" head) engines, the valve operating clearance is adjusted after the cylinder head is installed. The valve operating clearance for these engines is adjusted usually by turning the self-locking threaded nut that retains the valve rocker.

23. Thoroughly lubricate the valve stem and guide. Compress the intake valve spring (the lighter spring) and install the spring and retainer on the intake valve.

In similar manner install the exhaust valve spring (the heavier spring) and retainer. If a valve spring compressor is not available, the springs can be compressed in the bench vise and held with wire. Once the retainers are in position, cut the wire from the spring and remove it from the valve chamber.

Replace the valve cover using a new gasket. If the cover contains a crankcase breather, be certain to reassemble the unit in the proper sequence.

HEAD REPLACEMENT
AND FINAL REASSEMBLY

Figure 61

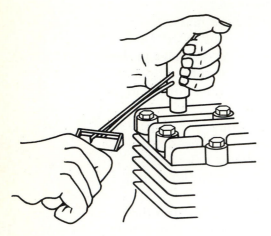

TIGHTEN HEAD BOLTS WITH A TORQUE WRENCH

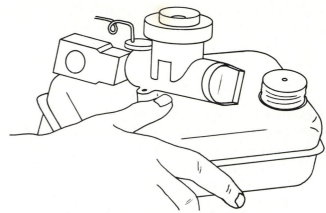

TILT THE FUEL TANK ASSEMBLY TO CONNECT THE CARBURETOR LINKAGE

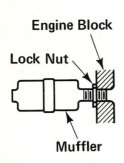

Engine Block

Lock Nut

Muffler

THREAD MUFFLER 1/2" INTO BLOCK AND SECURE WITH LOCKNUT

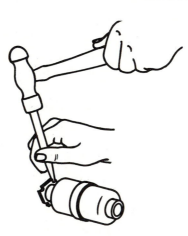

TIGHTEN MUFFLER LOCK NUT WITH HAMMER AND PUNCH

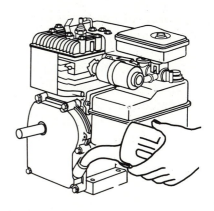

FILL THE CRANKCASE WITH SAE 30—SE or SF OIL

HEAD REPLACEMENT
AND FINAL REASSEMBLY

24. Make certain that the block and head are clean and place a new head gasket on the block. *Carefully* check the gasket to be sure that it aligns properly with the bolt holes and the block surface. Place a small amount of graphite grease on the threads of the head bolts and install the head on the engine. Make certain that all bolts are installed in the proper locations. If the bolts were mixed accidentally, check with a small rod or nail to determine where the longer bolts are needed to ensure that the bolts adequately thread into the engine.

25. Snug all head bolts by hand. Torque the bolts to one-third torque specifications following the tightening sequence shown for your bolt arrangement. Refer to the manufacturer's specifications for the torque value. Retighten all bolts to two-thirds torque specifications. Tighten to the torque specifications and go over them one more time to be certain that all bolts are at the proper torque. On "I" head engines, reassemble and adjust the valve train as required.

26. Adjust the gap and install a new spark plug. Torque the plug to the specified torque. If specifications are not available, torque the plug to 20 foot-pounds.

27. Replace the intake manifold and carburetor assembly. Connect the governor linkage before bolting the careburetor to the engine. (DO NOT BEND THE LINKAGE WIRE.)

28. Replace the ignition system and flywheel as outlined in the Ignition section.

29. Replace the muffler. *Caution:* On a model that has a muffler that threads into the engine block, DO NOT thread the muffler all the way into the block. Thread the muffler into the block approximately one-half inch and tighten the locking nut. If the muffler is threaded into the engine too far, it may "pinch" the exhaust valve guide or the tapered threads may distort the engine block.

30. Replace the air shroud and other components that were removed or disconnected.

31. Fill the crankcase with SAE 30, SE, or SF classification motor oil.
Note: Do not overfill the crankcase. Fill only to the full mark or level of the engine. After 5 hours of operation, change the engine oil. During break-in of a newly overhauled engine, the oil becomes contaminated quickly as the parts "wear in."

32. Replace the engine on the tiller, lawnmower, etc., and connect all controls.

33. Reconnect the drive train or blade.

34. Fill the fuel tank with clean fuel.

35. Start the engine and maintain a fast idle for the first few minutes of running.

36. Adjust the carburetor as outlined in the Carburetor section.

T F   1. Piston rings are fitted to the piston to keep the piston centered in the cylinder. Page 117.

T F   2. The crankshaft drives the camshaft at one-half crankshaft speed. Page 117.

T F   3. Gaskets and seals should not be reused when servicing the engine. Page 117.

T F   4. The part inside the engine that wears the most is the connecting rod bearing. Page 117.

T F   5. Failure to remove the rust from the drive end of the crankshaft can cause damage to the main bearing when the crankshaft is removed. Page 123.

T F   6. Any cylinder head that is warped must be replaced. Page 119.

T F   7. Some small-engine valve covers contain filter units for the crankcase breather. Page 125.

T F   8. Gasoline should *not* be used to clean the engine. Page 131.

T F   9. Cylinder head bolts used on small engines are all the same length. Page 119.

T F   10. Removing the carburetor requires bending the governor linkage. Page 123.

T F   11. The spark plug wire should be grounded before beginning work on the blade or drive unit. Page 123.

T F   12. Improper combustion can cause the piston rings to break or become ''locked'' in the ring groove. Page 137.

T F   13. On some engines the exhaust valve spring is heavier than the intake valve spring. Page 125.

T F   14. Identification marks should be painted on timing gears if there are no manufacturer marks. Page 129.

T F   15. All internal parts of the engine should be cleaned with soapy water when performing an overhaul. Page 131.

T F   16. A broken piston ring should never be used to clean the ring grooves of the piston. Page 137.

T F   17. If the cylinder of an engine is .015″ out of round, the engine should be junked. Page 133.

T F   18. Cylinder taper can be measured with a piston ring and feeler gauge. Page 133.

T F   19. A new piston should be installed in a cylinder that has been rebored. Page 137.

T F   20. Before removing the connecting rod from the piston, the rod and piston should be marked with a punch to aid in reassembly. Page 129.

T F   21. If the crankpin bearing surface of the connecting rod is scored, the connecting rod should be replaced. Page 141.

T F 22. The connecting rod cap should be filed down to achieve a proper fit on the crankshaft. Page 141.

T F 23. Most manufacturers recommend that bent crankshafts be straightened in a heavy press. Page 145.

T F 24. Wear on the camshaft is a common problem found in small engines. Page 147.

T F 25. On many small engines the crankshaft timing gear is made as a part of the crankshaft. Page 145.

T F 26. Crankshaft and camshaft timing gears wear quickly and must be replaced. Page 145.

T F 27. If the face of a valve lifter is badly scored, it should be refaced on a valve refacing machine. Page 149.

T F 28. A valve seat that is loose in the block should be welded in place with an arc welder. Page 151.

T F 29. The seating area of a valve seat should not be over 1/8" wide. Page 151.

T F 30. Valve margin should be at least 1/64". Page 153.

T F 31. When installing crankshaft seals, the sharp edge or lip should be toward the inside of the engine. Page 155.

T F 32. Piston ring end gap can be increased by filing the end of the ring. Page 159.

T F 33. The bevel on the inside edge of a piston ring should be toward the top of the piston. Page 159.

T F 34. A ring expander is used to hold the rings in place as the piston is installed in the cylinder. Page 159.

T F 35. The piston assembly should be hammered into the cylinder with a ball peen hammer. Page 161.

T F 36. Connecting rod bolts should be tightened with an open end wrench. Page 161.

T F 37. If the valve operating clearance is too much, the valve will have to be replaced unless the margin will allow additional refacing. Page 165.

T F 38. The engine oil should be changed after 5 hours of operation on a newly overhauled engine. Page 167.

39. Describe how the piston is attached to the connecting rod. Page 117.

40. How is the crankshaft supported in the engine block? Page 117.

41. List the functions of the piston rings. Page 117.

42. The connecting rod is attached to the _____ of the crankshaft. Page 117.

43. Identify the components and describe the operation of the valve train. Page 116.

44. Name two symptoms that would indicate an engine should be overhauled. Page 117.

45. What procedure should be followed in tightening cylinder head bolts? Page 119.

46. What four key points should be observed or practiced when overhauling an engine? Page 121.

47. Why must the ring ridge be remove from the cylinder as part of overhaul procedure? Page 125.

48. Describe how crankshaft end play can be measured. Page 127.

49. The crankshaft run-out or wobble should not exceed _____ . Page 127.

50. What tool should be used to remove connecting rod bolts or nuts? Page 129.

51. What does "10" mean when it is stamped on the top of the piston? Page 131.

52. Explain how carbon should be removed from the piston. Page 137.

53. What is cylinder taper? Page 133.
    Cylinder taper should not exceed _____ . Page 133.

54. List the three different types of cylinder wear that must be checked. Page 133.

55. Cylinder wear can be measured with a telescoping gauge and a _____ . Page 133.

56. If ring groove wear exceeds _____ , the piston should be replaced. Page 137.

57. Describe what can cause deep scores or scratches on the piston rings. Page 139.

58. How can piston pin wear be checked? Page 141.

59. Explain the difference between a plain-type bearing and an antifriction-type bearing. Page 143.

60. How can a worn valve guide be corrected or repaired? Page 149.

61. What parts are included in a new short block assembly? Page 147.

62. Identify the checks that should be made on antifriction bearings. Page 143.

63. Describe the measurments and checks that should be made on crankshaft journals. Page 143.

64. Explain the procedure for machining the valve seats in the engine block. Page 151.

65. Describe the procedure for replacing a crankshaft seal. Page 155.

66. Why is cleanliness important when reassembling an engine? Page 155.

67. Sketch an engine valve and label the main parts. Page 152.

68. Explain how crankshaft end play can be measured. Page 127.

69. Describe the procedure for checking piston ring end gap. Page 159.

70. Why should used piston rings *not* be reused in an overhauled engine? Page 159.

71. Identify by name the different rings used on the piston and the function of each. Page 159.

72. What precautions should be observed when installing piston rings on the piston? Page 159.

73. List the steps of procedure for checking and adjusting valve operating clearance. Page 165.

74. Explain how Plastiguage is used to check bearing clearance. Page 161.

75. What could cause excessive clearance between the connecting rod and the crankshaft? Page 161.

76. How can the crankcase cover be installed without damaging the crankshaft seal? Page 163.

Section IX

TWO-STROKE CYCLE ENGINE
SERVICE AND OVERHAUL

Figure 62

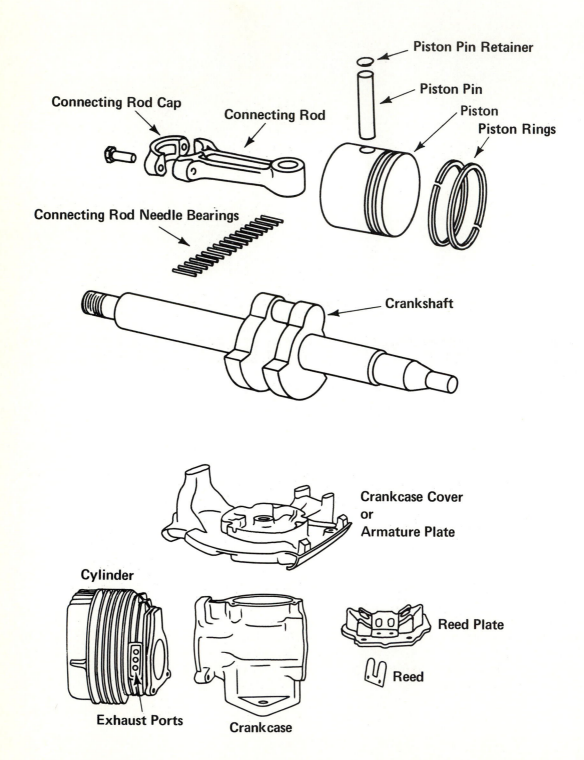

Connecting Rod Cap

Connecting Rod

Piston Pin Retainer

Piston Pin

Piston

Piston Rings

Connecting Rod Needle Bearings

Crankshaft

Crankcase Cover or Armature Plate

Cylinder

Reed Plate

Reed

Exhaust Ports

Crankcase

BASIC COMPONENTS

The two-stroke cycle engine delivers one power impulse each time the crankshaft completes one revolution. Intake and exhaust gases move in and out of the cylinder through ports (holes) in the sides of the lower part of the cylinder. This arrangement takes place of the valve train found on four-stroke cycle engines.

Oil is mixed with the fuel for the two-stroke cycle engine. Therefore, there is no need for a volume of oil in the crankcase. The oil suspended in the fuel vapor sticks to the surfaces of all the moving parts. This keeps all the parts coated with a film of oil, regardless of how much the engine is tilted.

Refer to the manufacturer's specifications on the label on the engine to determine the oil-to-gasoline ratio. The ratio varies, depending upon the age and manufacturer of the engine.

Reed valves (thin strips of metal) are located in the side of the crankcase. The reed valves permit the intake gases to enter the crankcase. The gases are transferred to the cylinder through the intake port in the lower part of the cylinder.

There is no exhaust valve used with the typical two-stroke cycle engine. The exhaust gases leave the cylinder through the exhaust port when the piston nears the end of the downward stroke.

The exhaust port tends to build up with carbon as a result of burning some of the oil in the fuel mixture. The port must be kept clean and free of this carbon.

The engine has main bearings and rod bearings that are similar to the four-stroke cycle engine. On some two-stroke engines, antifriction (roller) bearings are used. When servicing an engine with antifriction bearings, it is important that the rollers (needles) are not lost. Extreme care must be taken when disassembling the engine to make certain that all the bearing rollers are kept together.

The crankcase gaskets and crankshaft seals on the two-stroke cycle engine must be in good condition. Leakage past the seals or gaskets can prevent the engine from running properly. If the crankcase is not sealed completely, the intake gases will be diluted by the air leaked to the crankcase.

Figure 63

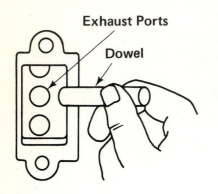

Exhaust Ports

Dowel

CLEANING CARBON FROM THE EXHAUST PORTS

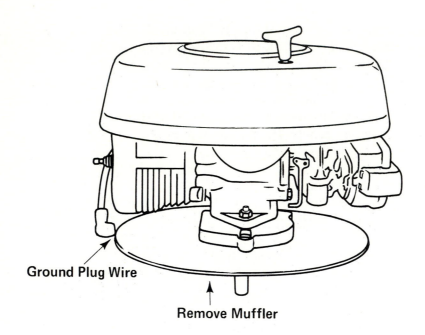

Ground Plug Wire

Remove Muffler

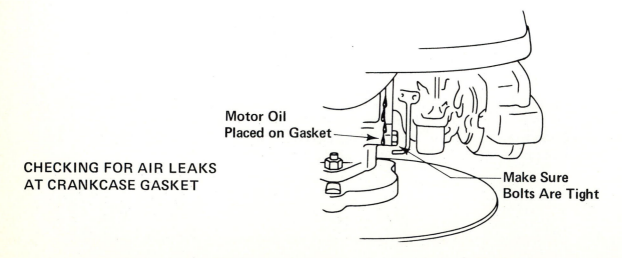

Motor Oil
Placed on Gasket

CHECKING FOR AIR LEAKS
AT CRANKCASE GASKET

Make Sure
Bolts Are Tight

MINOR ENGINE SERVICE

Two-stroke cycle engines require some basic services quite different from the four-stroke cycle engine. There is a tendency for carbon to build up around the ports of this engine because of the oil in the air-fuel mixture. If the carbon build up is excessive, the engine will lose power. *Caution:* Safety glasses should be worn when performing cleaning operations.

EXHAUST PORT CLEANING

1. Remove the shields and other components necessary to loosen the muffler or exhaust passage bolts.
2. Remove the muffler (or exhaust system).
3. Remove the spark plug and rotate the crankshaft until the piston blocks the exhaust port.   **Check Point** _____
4. Use a wooden dowel rod to scrape carbon from the ports. Be very careful that the cylinder walls and piston are not damaged.
5. WITH THE SPARK PLUG WIRE DISCONNECTED and connected to ground, crank the engine over to blow any carbon particles from the piston head and port area.
6. Clean the muffler and cover plate.   **Check Point** _____
7. Scrape the threads of the muffler bolts with a lead pencil. This will prevent the threads from seizing in the block casting.
8. Reassemble the muffler and shields.
9. Check the spark plug gap and replace the spark plug and spark plug wire.

CRANKCASE AIR LEAKS

The two-cycle engine depends on a sealed crankcase for efficient intake of gases into the crankcase and movement of the air-fuel mixture to the cylinder. Any defect that permits air to enter the crankcase on intake will also permit loss of the compressed air-fuel mixture in the crankcase at the time it is moved to the cylinder.

CHECKING FOR CRANKCASE AIR LEAKS

1. Check for loose screws and bolts that would allow crankcase leaks.
2. Tighten any loose components.
3. Apply a film of motor oil on the gasket sealing area of any portions that were loose.
4. Crank the engine and observe whether or not the oil is drawn into the crankcase through a defect in the gasket. Start the engine and continue to check suspicious areas for leaks. If any leakage is observed, the gasket should be replaced.
                                          Check Point _____

REED VALVE INSPECTION AND SERVICE

Figure 64

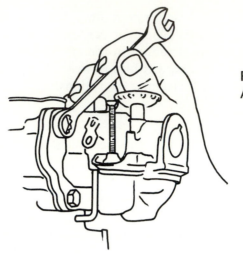

REMOVING CARBURETOR AND REED VALVE ASSEMBLY

INSTALL REED WITH SMOOTH EDGE TOWARD REED PLATE

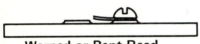

Warped or Bent Reed

Bent Convex

Bent Concave

Reed Plate

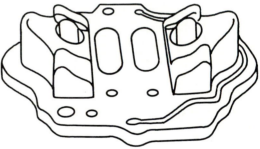

CAREFULLY INSPECT REED VALVES FOR THE PROBLEMS SHOWN ABOVE

Rough Edge → Smooth Edge

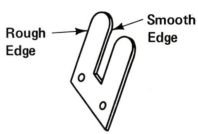

CHECK REED CLEARANCE WITH A FEELER GAUGE (REFER TO MANUFACTURER'S SPECIFICATIONS)

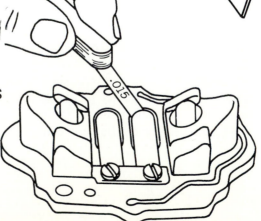

REED VALVE INSPECTION AND SERVICE

The reed valve(s) employed in the air-fuel intake system must function properly or else the two-stroke cycle engine will not run. A speck of dirt or a piece of grass in the reed valve will cause improper pressure buildup in the crankcase of the engine.

SERVICE PROCEDURE

1. Remove the carburetor assembly from the engine.
2. Carefully remove the reed valve assembly.
3. Inspect the unit for foreign material between the reed and its seat. Also inspect the reeds for warpage or other damage.
4. Remove the reed and rinse the valve components in clean solvent. AVOID USING COMPRESSED AIR because the blast of air can distort or bend the reeds. If the reeds are damaged or badly worn, replace them with new reeds.
5. Carefully reassemble the reed valve assembly. Check the clearance between the reed valve with a feeler gauge. Compare the clearance to specifications. If the clearance is improper, check to see that the unit is assembled properly.
    Note: Used reeds may appear to be good but not give the proper clearance. In such cases the reed should be replaced.

    Check Point _____
6. Using new gaskets, replace the reed valve assembly and carburetor on the engine.

ENGINE OVERHAUL
(TWO-STROKE CYCLE ENGINE)

Figure 65

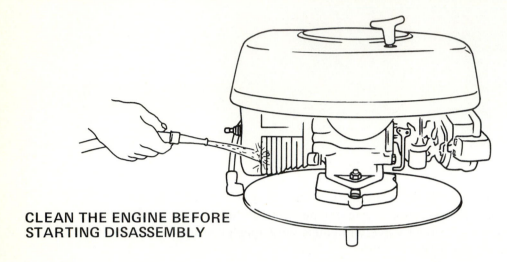

**CLEAN THE ENGINE BEFORE
STARTING DISASSEMBLY**

<u>CAUTION</u>: SAFETY GLASSES SHOULD BE WORN WHEN CLEANING THE ENGINE.

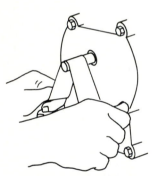

**POLISH DRIVE END OF
THE CRANKSHAFT TO
REMOVE RUST**

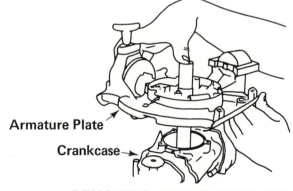

Armature Plate

Crankcase

**REMOVING ARMATURE PLATE
FROM THE CRANKCASE**

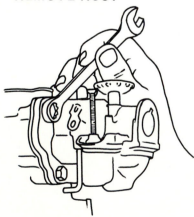

**REMOVING CARBURETOR
AND REED VALVE ASSEMBLY**

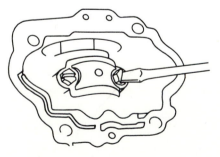

**LOOSENING CONNECTING
ROD BOLTS**

ENGINE OVERHAUL
(TWO-STROKE CYCLE ENGINE)

The engine should be disassembled completely for overhaul *only* if there is good reason. Thorough testing can determine whether or not an overhaul is necessary.

REASONS FOR OVERHAUL

1. Poor compression because of internal wear (compression testing is explained in the Compression Testing section of this text).
2. Excessive noise or knocks coming from inside the engine.
3. Crankshaft will not turn because of internal problem.

OVERHAUL PROCEDURE

1. Remove or disconnect the control cables and linkage. Disconnect the spark plug wire and attach it to the engine ground.
2. Disconnect the power drive belt, blade, or other power mechanisms. On shaft-driven, self-propelled units, mark the gears so that they can be reassembled in the same relative position.

Check Point _____

3. Remove the engine from the lawnmower, generator, or equipment.
4. Clean the outside of the engine. *Caution:* Safety glasses should be worn when performing cleaning operations! Hot, soapy water can be used to scrub the engine. Rinse with water.

On engines that are very dirty and greasy, apply engine cleaning solvent and rinse with water. DO NOT USE GASOLINE TO CLEAN THE ENGINE.

5. Dry the engine with cloths or compressed air.
6. Remove the shroud and the fuel tank. Note the types of screws used to mount these components. They should be replaced in the same positions.
7. Clean all rust and dirt from the drive end of the crankshaft as shown in the accompanying illustration. Any rust or roughness on the shaft will damage the drive-end main bearing when the crankshaft is removed from the crankcase.

Check Point _____

8. Check for a bent crankshaft. (See the Crankshaft area of the Four-Stroke Cycle Engine section for methods of checking the crankshaft.)
9. Remove the flywheel. On some models a filter screen must be removed to gain access to the flywheel nut. (See the Ignition section for the details on the flywheel removal.)
10. Remove the governor assembly and the armature plate. The armature plate contains a main bearing. On some engines the bearing is a loose-needle type. Be careful that none of the needles are lost.
11. Remove the carburetor and the reed plate.
12. Loosen the connecting rod bolts or nuts but DO NOT remove them from the rod.

CYLINDER, PISTON, AND CRANKSHAFT REMOVAL

Figure 66

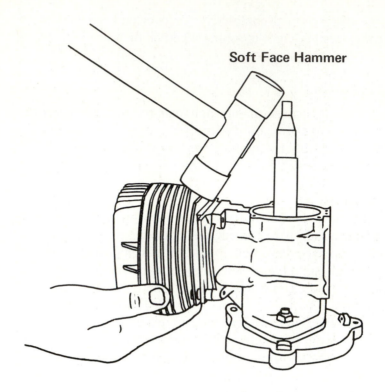

Soft Face Hammer

REMOVING CYLINDER (JUG) FROM THE CRANKCASE

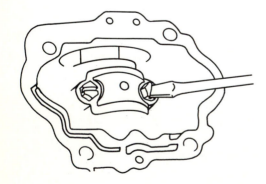

REMOVING CONNECTING ROD CAP

REMOVING THE CRANKSHAFT

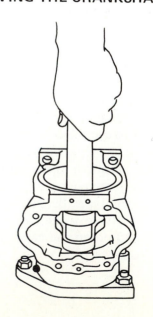

182

CYLINDER, PISTON, AND CRANKSHAFT REMOVAL

13. Remove the bolts or screws that attach the cylinder (jug) or the cylinder head. *Note:* There is no provision for removing the cylinder head or cylinder on some two-cycle engines. On these engines the piston is removed through the bottom of the cylinder.

14. Tap the cylinder with a soft hammer to break it loose from the crankcase.

15. Remove the cylinder by pulling it away from the piston quickly. Note the match marks on the connecting rod and the connecting rod cap. Remove the connecting rod bolts or nuts and remove the piston assembly from the engine. Be careful to retain all the loose-needle bearings of the connecting rod bearing. *Note:* On some two-cycle engines the rod *cannot* be removed from the crankshaft. On such engines check carefully for roughness by revolving the connecting rod on the crankshaft. If there is roughness or if the side play is excessive, the crankshaft-rod assembly should be replaced. The costs for this type of repair should be weighed against the costs for replacing the entire engine.

16. Remove the crankshaft from the crankcase.

CHECKING WEAR
OF ENGINE COMPONENTS

Figure 67

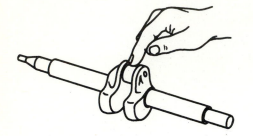

FEEL FOR WAVINESS ON THE BEARING JOURNAL ON ENGINES WHICH HAVE ROLLER BEARINGS

POINTS OF MEASUREMENT FOR CHECKING JOURNAL OUT-OF-ROUND

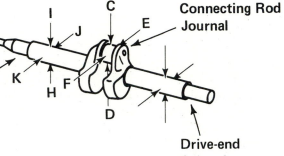

Connecting Rod Journal

Flywheel or Magneto-end Journal

Drive-end Journal

Scored Cylinder

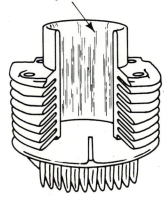

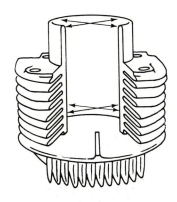

EXAMINE THE CYLINDER FOR SCORING

MEASUREMENT POINTS FOR CHECKING CYLINDER WEAR

NOTES: _____

_____

_____

_____

_____

_____

_____

CHECKING WEAR
OF ENGINE COMPONENTS

17. Wipe crankshaft bearing journals with a clean cloth. Carefully inspect the journals that employ needle-type bearings. Wear can be noted by feeling the journal. Slight roughness or waviness indicates wear and the crankshaft should be replaced.   **Check Point** _____

When the crankshaft journal is a plain bearing, it should be measured with a micrometer. Refer to the Crankshaft Measurement section in Four-Stroke Cycle Overhaul for information on measuring crankshaft journals. Compare the measurements to the manufacturer's specifications. If they exceed the maximum acceptable wear, the crankshaft should be replaced. If the manufacturer's data are not available, .001″ can be considered the maximum out-of-round which is acceptable. Record the crankshaft out-of-round and taper in the Data Block.

18. Inspect the cylinder for wear and scoring. Some two-cycle engines utilize a chrome-plate cylinder. If the chrome has "worn through," the cylinder should be replaced. If the cylinder is scored, it should be replaced.

Check the cylinder for wear and taper by taking measurements with a telescoping gauge and micrometer. For additional information on cylinder measurements, refer to the Cylinder Measurement section in Four-Stroke Cycle Overhaul. Compare the measurements to determine out-of-round and taper. Check the manufacturer's specifications for the maximum allowable taper and wear. Record the information in the Data Block.

If the wear or taper exceeds the maximum, the cylinder should be bored or honed oversize. The cylinder resizing should be done by a machine shop equipped to perform this machining. Check the availability of an oversize piston and rings before reconditioning the cylinder. In some instances it may be more economical to replace the cylinder rather than have it reconditioned.

Data Block

	Actual	Specifications
Drive-end main journal out-of-round	_____	_____
Magneto-end main journal out-of-round	_____	_____
Connecting rod journal out-of-round	_____	_____
Drive-end main journal taper	_____	_____
Magneto-end journal taper	_____	_____
Connecting rod taper	_____	_____
Cylinder wear	_____	_____
Cylinder out-of-round	_____	_____
Cylinder taper	_____	_____

PISTON PIN AND CONNECTING ROD CHECKS

Figure 68

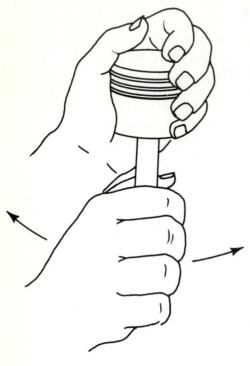

CHECKING FOR
PISTON PIN WEAR

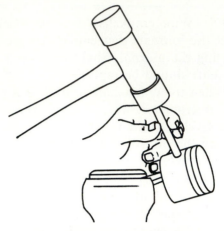

REMOVING PISTON PIN

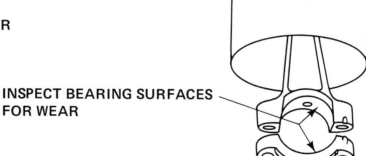

INSPECT BEARING SURFACES
FOR WEAR

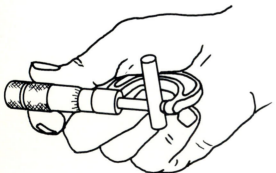

CHECKING PISTON PIN WEAR

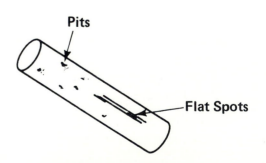

Pits

Flat Spots

CAREFULLY EXAMINE
BEARING ROLLERS

PISTON PIN AND CONNECTING ROD CHECKS

19. Piston pin wear can be checked by holding the piston firmly in one hand while attempting to "rock" the rod with the other hand. The rod must be "rocked" in line with the piston pin. It is normal for the rod to slide freely across the piston pin. Do not confuse this with piston pin wear.

If looseness is detected, the piston pin should be removed. Before removing the pin, note identifying marks or mark the piston and rod with a punch mark so that they can be reassembled properly. Remove the piston pin retainer with small nose pliers and remove the piston pin. Be careful that none of the rollers is lost on units which utilize a needle-type bearing.

Examine the piston pin and connecting rod pin bore for scoring or roughness. Any component that is scored should be replaced.

Measure the piston pin and the connecting rod bore with a micrometer.

On a model that utilizes a plain-type bearing at the pin, the clearance between the piston pin and the connecting rod pin bore should not exceed .002" (.051 mm) on most engines.

Oversize pins are available from some manufacturers. When an oversize pin is being installed, the rod and piston must be reamed or honed to achieve the proper fit.

Check Point _____

20. Inspect the connecting rod bearing for scoring and wear. The bore can be measured with a telescoping gauge and micrometer to find its size. Compare the measurement to the manufacturer's specification to determine wear. If over .0005" (.0127 mm) wear is present, the rod should be replaced.

On a model that utilizes a needle bearing in the rod, carefully examine the rod or the insert for roughness. If roughness is present, replace the component. Inspect the rod bearing needles for pits and flat spots. If either is present, the complete set of needles should be replaced.

Check Point _____

PISTON CHECKS

Figure 69

Scored Piston

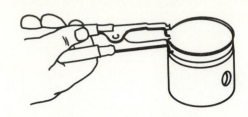

REMOVING PISTON RINGS WITH A
PISTON RING EXPANDER

CAUTION: SAFETY GLASSES SHOULD
BE WORN WHEN CLEANING
THE PISTON.

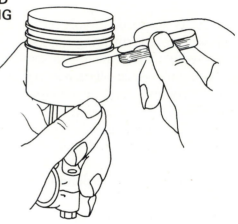

CHECKING PISTON RING GROOVE WEAR

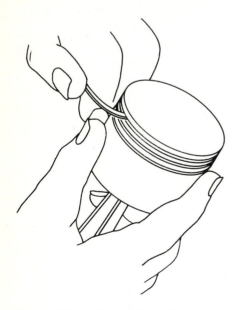

CLEANING CARBON FROM THE
RING GROOVES WITH A
BROKEN PISTON RING

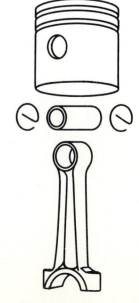

PREPARING TO INSTALL THE
CONNECTING ROD ON
THE PISTON

PISTON CHECKS

21. Carefully examine the piston for scoring and wear. If a two-cycle engine is operated on fuel with no oil added, the piston will score very quickly. If too much oil or a low-grade oil is added to the fuel, carbon will build up quickly in the exhaust port. This excessive carbon can cause scoring of the piston.

A badly scored piston should be replaced. To verify the wear of a piston, measure it with a micrometer and compare the measurements to the manufacturer's specifications.

22. If the piston is not scored or worn, it can be reused. Remove the piston rings with a piston ring expander. The piston ring grooves must be cleaned thoroughly. Many manufacturers use "pins" or wires to prevent the piston rings from turning on the piston. Either remove the pins or wires or use extreme caution when cleaning the ring grooves. Clean all carbon from the grooves with a piston ring groove cleaner or a broken piston ring. If a broken ring is used, grind the end of the ring square so that the carbon in the base of the groove can be removed.

23. Check the ring groove wear by placing a new ring in the groove and measuring the side clearance with a feeler gauge. The side clearance measurement should be checked against the manufacturer's specifications. If the manufacturer's data are not available, .006 (.0152 mm) is considered maximum side clearance on many engines. If the ring side clearance is excessive, the piston should be replaced.

Check Point _____

24. Install the piston on the connecting rod. (Match the punch or identifying marks on the rod and the piston.) Be certain that the rod and piston are assembled so that when installed in the engine the long slope of the piston is on the exhaust port side of the cylinder. When installing the piston pin retainers, make certain that the retainer ring opening is toward the top of the piston.

PISTON RING CHECKS AND INSTALLATION

Figure 70

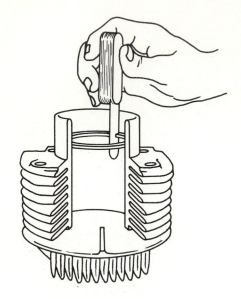

CHECKING THE PISTON RING END GAP CLEARANCE WITH A FEELER GAUGE

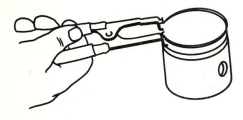

INSTALLING PISTON RINGS WITH A PISTON RING EXPANDER

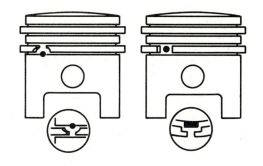

METHODS OF PINNING PISTON RINGS

Side View

Pin   Ring

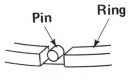

Top View

Ring   Pin

POSITION RING ENDS AROUND THE PIN RETAINER

PISTON RING CHECKS AND INSTALLATION

Cleanliness is most important when reassembling an engine. Make certain that the work area, tools, component parts, and your hands are clean before beginning the reassembly of the engine.

25. Clean all engine parts in solvent. Caution: Safety glasses should be worn when performing cleaning operations! Blow the excess solvent from the parts with compressed air and lay each part on a clean work surface.

26. Check the ring end gap of the new piston rings in the cylinder. ALWAYS INSTALL NEW PISTON RINGS. Position the ring squarely in the cylinder and measure the gap between the ring ends with a feeler gauge. Check the manufacturer's data for acceptable end gap. If these data are not available, .005″ (.0127 mm) is considered the minimum gap. If the gap is less than the minimum or .005″ (.0127 mm), check to be certain that the rings are the proper ones for the engine. On some models the end gap can be filed with a file to achieve proper clearance. Generally the gap should not exceed .020″ (.508 mm).

27. Carefully install the rings on the piston. Safety glasses should be worn when installing piston rings. Use a piston ring expander for this operation. If no ring expander is available, carefully place the rings on the piston. Install the lower piston ring first. If the piston has pin or wire retainers, be careful to position the ring properly over the retainer. The pin retainers or wire prevent the ring from rotating to a position where the end of the ring would catch in the cylinder port.

Check Point _____

SEAL REPLACEMENT AND CRANKSHAFT INSTALLATION

Figure 71

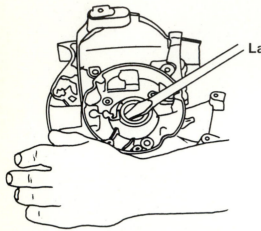

Large Screwdriver

REMOVING CRANKSHAFT SEALS

CAUTION: SAFETY GLASSES SHOULD BE WORN WHEN PERFORMING THIS OPERATION.

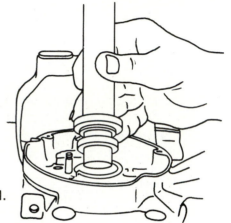

INSTALL NEW CRANKSHAFT SEALS WITH THE LIP OF THE SEAL TOWARD THE INSIDE OF THE ENGINE

INSTALLING THE CRANKSHAFT. (ON ENGINES WITH LOOSE ROLLER BEARINGS, COAT THE ROLLERS WITH GREASE TO KEEP THEM IN PLACE.)

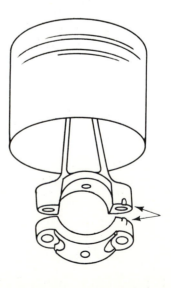

POSITION THE CONNECTING ROD PROPERLY ON THE CRANKSHAFT

SEAL REPLACEMENT AND CRANKSHAFT INSTALLATION

28. Remove the old seals from the crankcase and/or crankcase covers with a large screwdriver.

29. Install new crankshaft seals with a seal installation tool or a block of wood. Be certain that the lip of the seal is toward the inside of the engine. *Note:* On some engines that employ needle-type main bearings, the seal is installed after the bearings are in place in the crankcase and armature plate.

30. Wipe the main bearing of the block and lubricate it with SAE 20, SE, or SF oil. Wipe the crankshaft clean, oil with clean motor oil, and install it in the crankcase. BE CAREFUL that the oil seal is not damaged as the shaft is installed. On models with loose-needle bearings, coat the needles with Vaseline or other low-melting temperature grease to keep the rollers in place while the crankshaft is being installed.

31. Position the piston assembly in place. Be certain that the piston is positioned identical to its position before disassembly. If a rod "liner" is used, position the liner in the connecting rod. Place one-half of the loose needles in the connecting rod. Keep the needles in place with grease and pull the rod into place on the crankshaft. Position the remaining needle bearings in the cap or on the crankshaft and install the connecting rod cap.

Make certain that the match marks on the rod and rod cap are aligned. Carefully tighten the rod bolts until they are *snug*. DO NOT TIGHTEN THE CONNECTING ROD BOLTS AT THIS TIME.

Note: On engines with a plain-type rod bearing, disregard the above needle-bearing instructions.

Check Point _____

CYLINDER AND CRANKCASE REASSEMBLY

Figure 72

CAUTION: SAFETY GLASSES SHOULD BE WORN WHEN PERFORMING THE ENGINE REASSEMBLY OPERATIONS!

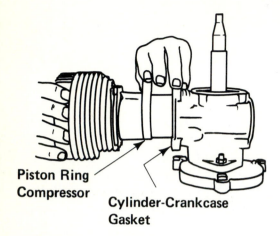

Piston Ring Compressor

Cylinder-Crankcase Gasket

INSTALLING THE CYLINDER (JUG) OVER THE PISTON

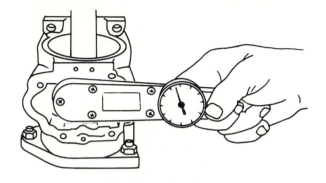

TIGHTEN THE CONNECTING ROD BOLTS TO SPECIFICATIONS WITH A TORQUE WRENCH

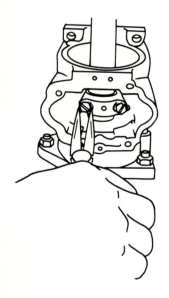

BEND THE LOCK TABS AGAINST THE HEADS OF THE CONNECTING

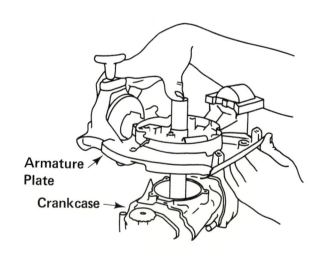

Armature Plate

Crankcase

REPLACING THE ARMATURE PLATE

CYLINDER AND CRANKCASE REASSEMBLY

32. Place the cylinder-crankcase gasket on the crankcase. *Be certain that the port holes are properly aligned.*

33. Apply a liberal coat of oil to the piston and piston rings. Position the rings properly over the retainers on pistons so designed. On engines without pins, stagger the rings so that the end gaps are not in line with each other and are not in alignment with the cylinder ports. Install a ring compressor over the rings. Wipe the cylinder clean, apply a liberal coating of oil, and slide the cylinder (jug) over the piston.

34. Tighten the connecting rod bolts to the proper torque specifications with a torque wrench. Bend the lock tabs around the rod bolts or nuts. Rotate the crankshaft to be certain that there is no binding.

35. Install the armature plate or crankcase cover wtih a new gasket between the cover and the crankcase. Apply grease to the needle bearings to hold them in place if this type bearing is employed. *Be careful* that the new crankshaft seal is not damaged when sliding the cover into place. A film of wax paper can be wrapped around the crankshaft to prevent the seal from catching on the main bearing shoulder. Tighten the cover retaining bolts.

36. Install the reed valve assembly and carburetor. Use new gaskets. Be certain that the reeds are toward the engine! Tighten the reed valve and carburetor retaining nuts or bolts.

37. Install the ignition system components and flywheel. See the Ignition Service section for details on the ignition system.

FINAL STEPS OF REASSEMBLY

Figure 73

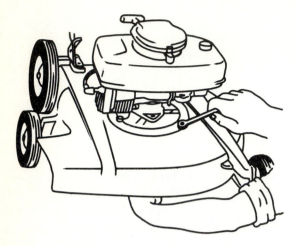

INSTALLING THE ENGINE

REPLACING SHROUDS AND
EXTERNAL COMPONENTS

MIX GASOLINE AND "2-CYCLE"
OIL IN THE CORRECT RATIO.
FOLLOW THE MANUFACTURER'S
RECOMMENDED PROPORTION.

CAUTION: FUEL SHOULD BE POURED
INTO THE TANK OUTDOORS!

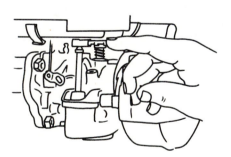

ADJUSTING THE CARBURETOR

FINAL STEPS OF
REASSEMBLY

38.   Replace the flywheel shrouds. Reconnect the carburetor linkages and the fuel lines.

39.   Reassemble the engine on the lawnmower or powered unit and reconnect all the control cables and mechanisms.

40.   Fill the fuel tank with the proper mixture of fuel and oil. Caution: Fuel should be added outdoors!

41.   Perform the basic carburetor adjustments as outlined in the Carburetor Service section.

42.   Start the engine and make final carburetor adjustments. Caution: Exercise caution around drive mechanisms, belts and blades, or cutters!

Check Point _____

43.   Operate the engine under normal conditions and readjust the carburetor as is needed.

Questions for Section IX

T F 1. The two-stroke cycle engine delivers a power impulse each time the piston moves to the top of the cylinder. Page 175.

T F 2. Ports in the two-stroke cycle engine serve the same function as the valves in a four-stroke cycle engine. Page 175.

T F 3. Fuel oil is mixed with the gasoline for most two-stroke cycle engines. Page 175.

T F 4. The spark plug wire should be attached to a good ground before starting to work on the blade or drive mechanism. Page 177.

T F 5. The cylinder ports must be periodically scraped with a round file. Page 177.

T F 6. A speck of dirt or grass can cause a reed valve to leak. Page 179.

T F 7. Before starting the overhaul the engine should be thoroughly cleaned with gasoline. Page 181.

T F 8. If the connecting rod journal of the crankshaft feels "wavy," over-sized bearing rollers should be installed. Page 185.

T F 9. On some engines the connecting rod should not be removed from the crankshaft. Page 183.

T F 10. Generally speaking, crankshaft out-of-round should not be more than .001 " (.025 mm). Page 185.

T F 11. Some rod bearings use loose rollers that fall out when the connecting rod cap is removed. Page 183.

T F 12. A badly scored piston should be ground down and resurfaced. Page 189.

T F 13. Waxpaper can be wrapped around the crankshaft to prevent damage to the seal when assembling the engine. Page 195.

14. How do the ports control the flow of intake and exhaust gasses? Page 175.

15. Explain the operation of the reed valve. Page 175.

16. List some of the reasons for overhauling a two-stroke cycle engine. Page 181.

17. What are some of the problems that occur with reed valves? Page 179.

18. How can crankcase "leaks" be found? Page 177.

19. Why must the crankcase seals be in good condition? Page 177.

20. What damage can occur if the rust is not removed from the drive end of the crankshaft on an engine that is being overhauled? Page 181.

21. Describe the checks that can be made to reveal a bent crankshaft Page 181.

22. Explain the checks that should be made on the crankshaft of an engine that is being overhauled. Page 185.

23. List the checks that should be made of the cylinder of an engine that is being overhauled. Page 185.

24. How can piston pin wear be checked? Page 187.

25. Connecting rod needle bearings should be checked for _____ . Page 187.

26. Identify the precautions to be observed when installing the connecting rod cap. Page 193.

27. Why are pins or wires used to locate the piston rings on the piston on a two-stroke cycle engine? Page 190.

28. List some reasons for a scratched or scored piston. Page 189.

29. The maximum side clearance of a piston ring is. _____ . Page 189.

30. What tool is used to install piston rings on the piston? Page 190.

31. Why should new seals be installed in the crankcase when the engine is overhauled? Page 175.

32. Explain how the reed valve assembly should be installed on the engine. Page 195.

33. How can loose needle bearings be held in place when assembling the engine? Page 193.

THE STORAGE BATTERY

Section X
BATTERY AND STARTER TESTING

THE STORAGE BATTERY

Figure 74

CAUTION: SAFETY GLASSES SHOULD BE WORN AND JEWELRY
(RINGS AND WATCHES) SHOULD BE REMOVED
WHEN PERFORMING ELECTRICAL TESTS!

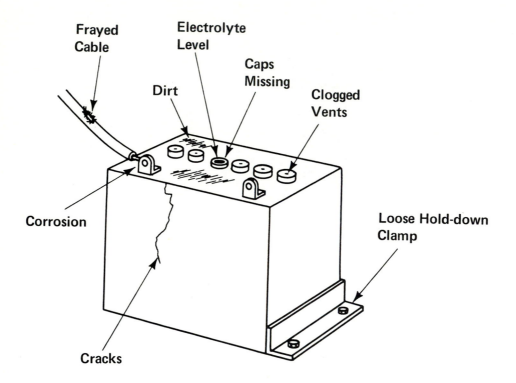

THE STORAGE BATTERY

The storage battery is designed to chemically accept an electrical charge and hold it until needed. Batteries have a number of characteristics and weaknesses that are usually overlooked until it is too late. This is especially true on lawn and garden equipment in which the entire charging and starting circuits may be only accessories. The location of the battery is not a primary design consideration and is usually not located to the battery's best advantage.

Electrolyte. Electrical power is stored chemically in the battery. The chemical balance of the battery must be maintained. Do not add chemicals to the battery because they will only destroy the chemical balance.

Do not dump a battery and refill it with clean water or acid. As the battery takes on a charge, the chemical composition of the electrolyte solution changes. The composition of the electrolyte is determined by the state of charge of the battery. Replacing the old electrolyte with new would upset the battery's chemical balance.

Heat and overcharging are the usual causes of electrolyte loss. The loss is water. Replace the water. Do not overfill or allow the battery to operate with a low electrolyte level.

Use clean water to refill the battery because there is only a limited area below the plates for foreign matter to collect. Once the area below the plates becomes filled, the cell becomes shorted and the battery is ruined. *Note:* Some batteries lose very little electrolyte and therefore *do not* have vent caps.

CAUTION: While the battery is being charged it is giving off an explosive gas through the vents in the caps. Do not smoke around a battery that is being charged or has recently been charged. This applies to batteries on equipment that has been running because the battery was being charged while the engine was running. Always turn off all accessories and battery chargers before removing clamps and cables from the battery terminals to prevent sparks that might ignite fumes still collected near the vent caps.

Visual Check. Batteries should be mounted firmly on the equipment to prevent overturning and short circuits if the battery moves around. Think what might happen if you were to turn over a riding lawnmower and the battery were not secure. You could be covered with acid or fire or an explosion could occur.

Cracks in the case or defects in the sealing compound indicate battery faults and should be corrected or replaced. Some battery defects can be repaired.

BATTERY TESTS

Figure 75

CAUTION: SAFETY GLASSES SHOULD BE WORN AND JEWELRY (RINGS AND WATCHES) SHOULD BE REMOVED WHEN TESTING OR SERVICING BATTERIES!

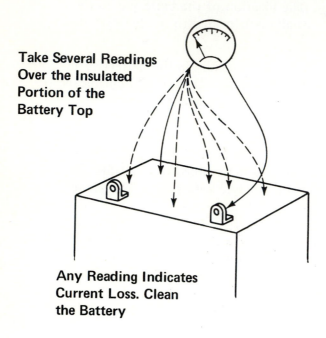

Take Several Readings Over the Insulated Portion of the Battery Top

Any Reading Indicates Current Loss. Clean the Battery

TOOLS NEEDED:

1. D.C. Voltmeter, 16 Volts or More. 25-Volt Scale Is Desirable.
2. Baking Soda Solution and Brush.
3. Battery Hydrometer

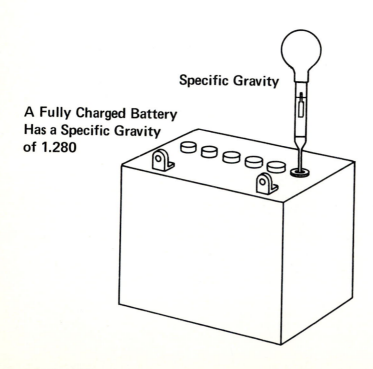

Specific Gravity

A Fully Charged Battery Has a Specific Gravity of 1.280

BATTERY TESTS

Several tests may be performed to determine the condition of the battery. Any one test will not ensure a complete analysis. Making an assumption about a cause of a problem based on only one test would be like a doctor performing brain surgery on you because you have a headache when perhaps an aspirin would cure it. Make as many tests as possible to determine the battery condition.

SELF-DISCHARGE

A common problem is dirt and moisture buildup on top of the battery, which creates a path for current flow or discharge across the top of the battery. Check for this by connecting the negative voltmeter lead to the battery and touching the positive voltmeter probe several places on the insulated portion of the battery top. Any voltage found here is a result of leakage of current. Clean the battery top. Use a baking soda solution to help remove corrosion and acid deposits. Do not allow the baking soda solution to enter the battery. *Always wear safety glasses and wash your hands frequently when you work with a battery. Wash off any liquid that splashes on your face or arms.*

SPECIFIC GRAVITY TEST

As the battery is charged and discharged, both chemical solution and its specific gravity (weight) are changed. The charge of each cell can be measured with a battery *hydrometer*. Hydrometers are calibrated to read correctly at 80° or room temperature. If the test must be made at any other temperature, a hydrometer with a temperature correction scale must be used to get an accurate reading.

Draw enough water into the hydrometer to allow the bulb to float freely and to record the reading accurately. Perform this test on each cell of the battery and compare the readings. A fully charged battery should read near 1.280. These readings indicate only the state of charge the battery currently has placed on it. A perfectly good battery will read low if the charging system is not keeping it charged or if considerable cranking or accessory load has just been placed on it.

One really important indication is that whatever the state of charge of the battery, all cells should read the same. The cells are connected in series and the charge and discharge of each cell are equal. Any variation between the readings obtained from the cells indicates bad cells and a weakened battery. If this battery is continued in use, it will place extra load on the charging circuit and will have reduced cranking power.

BATTERY CAPACITY TESTS

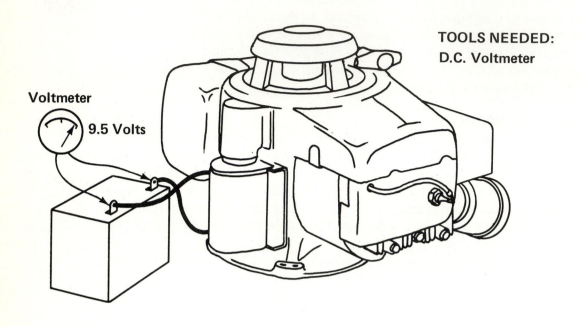

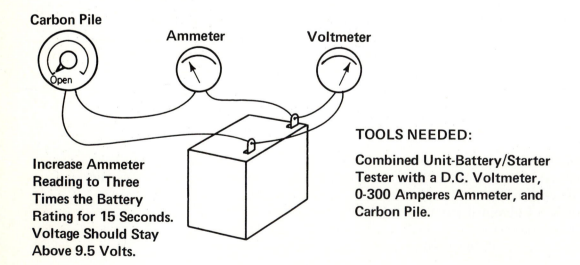

Figure 76

TOOLS NEEDED:
D.C. Voltmeter

Voltmeter

9.5 Volts

CAUTION: SAFETY GLASSES SHOULD BE WORN AND JEWELRY (RINGS AND WATCHES) SHOULD BE REMOVED WHEN TESTING OR SERVICING BATTERIES!

Carbon Pile

Open

Ammeter

Voltmeter

Increase Ammeter
Reading to Three
Times the Battery
Rating for 15 Seconds.
Voltage Should Stay
Above 9.5 Volts.

TOOLS NEEDED:

Combined Unit-Battery/Starter
Tester with a D.C. Voltmeter,
0-300 Amperes Ammeter, and
Carbon Pile.

BATTERY CAPACITY TESTS

Gravity tests cannot be made on batteries that do not have removable cell caps. On these batteries, testing can be done as outlined in the following procedures.

The battery must be able to maintain voltage during the cranking operation. Just taking a voltage reading from a battery not in use does not tell much. A nearly dead battery can show full voltage (12.6 volts D.C.) when no load is connected. The true test of capacity is the ability to maintain voltage while under load.

A simple battery test can be made by connecting a voltmeter across the battery while the battery is still on the equipment. The no-load reading should be 12.6 volts. Observe the voltmeter and crank the engine. Crank until the voltmeter drops to a steady hold position. *Caution:* Do not crank longer than 15 seconds! The voltmeter reading should be above 9.5 volts. If the engine cranked at a good normal speed and if the voltage remained above 9.5 volts, the battery is satisfactory. Failure to pass this test could be caused by excessive starter draw. Check the battery by using the carbon pile method.

CARBON PILE TEST

A *carbon pile* is a variable load device. Connect the carbon pile in series with an ammeter. The ammeter will indicate the amount of load being placed on the battery by the carbon pile. Connect a voltmeter across the battery to measure the battery voltage. Adjust the carbon pile to three times the ampere-hour rating of the battery for 15 seconds and observe the voltmeter reading. Do not leave the load on the battery for more than 15 seconds because the battery will be quickly discharged. If the battery reading does not drop below 9.5 volts, the battery will provide dependable performance. If the battery drops below 9.0 volts, it is either defective or is not fully charged. Retest with a hydrometer, and recharge the battery if necessary. If a fully charged battery does not pass the test, it is defective and must be replaced.

EQUIPMENT

The ammeter and carbon pile used for this test must be able to handle three times the ampere-hour rating of the battery being tested. Battery/starter testers with a D.C. voltmeter, suitable ammeter, and carbon pile are available.

TEST RESULTS

Figure 77

CAUTION: SAFETY GLASSES SHOULD BE WORN AND JEWELRY (RINGS AND WATCHES) SHOULD BE REMOVED WHEN TESTING OR SERVICING BATTERIES!

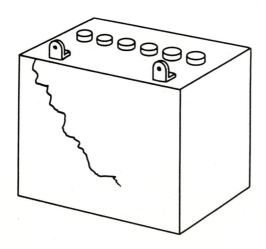

TEST RESULTS

TEST	INDICATION	SOLUTION
Visual	Loose hold-down clamp.	Tighten clamp or replace with new one if necessary.
	Corrosion on battery posts and clamps.	Clean with baking soda solution.
	Top of battery wet.	Possible overcharge. Check charging circuit.
	Frayed cables.	Repair or replace.
	Cracked case.	Repair or replace battery.
	Electrolyte level low.	Refill with clean water. Could be caused by over-charging. Check charging voltage.
Self-discharge	Readings found across top of battery.	Clean battery with baking soda solution.
Hydrometer	All cells show 1.280 specific gravity.	Fully charge battery.
	All cells show 1.280 except one.	Bad cell-battery will fail soon.
	All cells read low but equal.	Battery probably OK. Needs charging.
Cranking capacity test	Voltage remained above 9.5 volts during crank-ing—engine cranked normally.	Battery OK.
	Voltage dropped below 9.0 volts.	Battery defective or not fully charged. Check battery charge with hydrometer. Recharge and retest.
	Battery recharged, but voltage still drops below 9.0 volts during cranking.	Battery defective or starter circuit problem. Check battery with carbon pile.
	Battery stays above 9.5 volts, but starter slow or not working at all.	Check starter circuit. See page 215–217.
Carbon pile test	Battery voltage remains above 9.5 volts while am-meter reads three times battery ampere-hour rating.	Battery OK.
	Battery voltage drops below 9.0 volts.	Battery defective if fully charged. Check with hydrometer and retest.

209

Figure 78

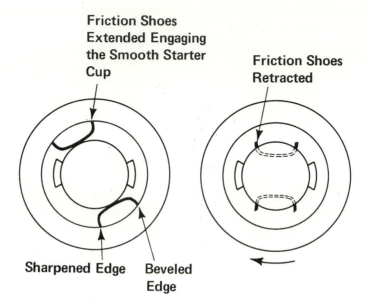

Friction Shoes
Extended Engaging
the Smooth Starter
Cup

Friction Shoes
Retracted

Sharpened Edge   Beveled
Edge

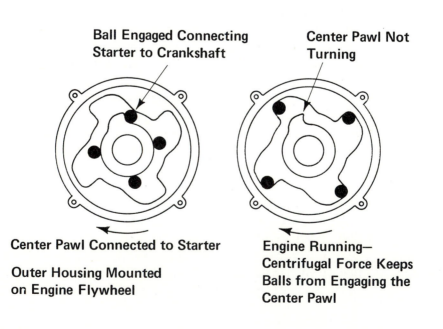

Ball Engaged Connecting
Starter to Crankshaft

Center Pawl Not
Turning

Center Pawl Connected to Starter

Outer Housing Mounted
on Engine Flywheel

Engine Running—
Centrifugal Force Keeps
Balls from Engaging the
Center Pawl

REWIND STARTERS

A number of rewind starter types are used on small gas engines. Regardless of the type, they are subject to the same wear and abuse. Basically, they consist of a pull rope, recoil spring, and an engaging machanism. The variations occur in the engaging mechanism. These mechanisms should be basically understood so that wear and mechanical failures can be determined. Since repair usually consists of just replacing the worn parts, detailed repair instructions are not necessary.

FRICTION SHOE TYPE

The friction shoe type starter utilizes cam action to engage the friction shoe against a smooth starter cup. As with the dog type, a brake spring creates a slight drag on the shoe unit holding the shoes from turning until the cam under the shoes moves, pushing the shoes out against the starter cup engaging the starter rope with the flywheel.

BALL TYPE

When the engine is stopped, the balls roll down the inclined ramp to the center pawl. When the rope is pulled, the center pawl catches one of the balls and locks it against the starter cup engaging the pawl with the starter cup. As soon as the engine starts, centrifugal force carries the balls out into the recesses and they continue to turn with the engine. The pawl remains stopped with the rewind rope. This type screws onto the crankshaft and requires a special tool (see Flywheel Removal) to remove. The ears cast on the starter cup are easily broken off if a hammer is used to loosen the starter assembly. The entire assembly slides on over the extended crankshaft and is removed as an assembly. On older models the unit is held together by a large snap ring that is easily pried out. Newer units are sealed. The sealed cover may be removed by using a small chisel to pry off the retainer cover. Do not use a screwdriver because it will be necessary to strike it with a hammer to remove the cover and most screwdrivers were not intended to be struck with a hammer and they made explode.

When replacing the ball-type starter unit, be sure that it slips freely over the extended crankshaft end because the crankshaft must turn inside the starter pawl. Failure of the shaft to turn freely inside the pawl will cause the starter to "catch" occasionally, making a loud scratching sound. Polish the shaft as needed to remove burrs and rust so that the assembly slides freely on the shaft.

Figure 79

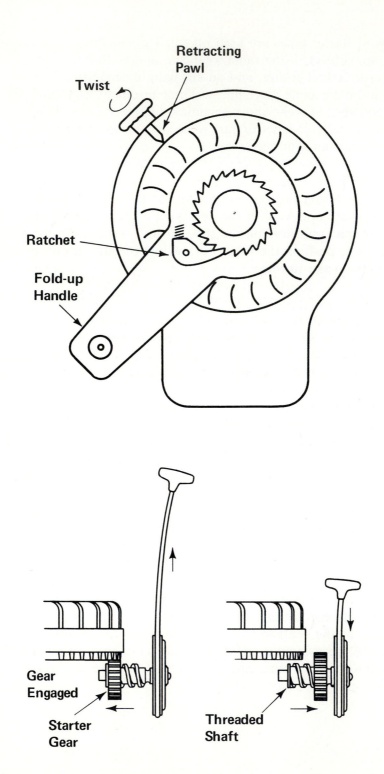

IMPULSE WINDUP TYPE

The windup starter uses one of the clutches described previously, but it allows the spring to be wound up tight while the flywheel is held locked in place by a locking pawl. A fold-up handle ratchets around the crankshaft gear while the spring is being wound up. Once the spring is securely wound, the lock is released, thus allowing the spring to crank the engine. On older types the lock pawl tended to wear a groove in the aluminum flywheel, resulting in partial "releases" of the flywheel while trying to wind the spring. Remounting the lock mechanism so that it engages the flywheel at a different point is sometimes possible or the flywheel may be replaced. The entire starter assembly and blower housing can easily and inexpensively be replaced with the conventional recoil starter. See the manufacturer's literature for spring rewinding instructions for that particular model starter.

VERTICAL PULL STARTER

The vertical pull starter utilizes a starter gear that engages a gear ring on the flywheel. The small starter gear is fitted on a threaded shaft. The first partial pull of the starter rope causes the starter gear to move down the threaded shaft to engage the flywheel ring gear. As the recoil spring rewinds the rope, the gear moves back down the threaded shaft disengaging it from the flywheel. When the engine starts, the flywheel spins the small starter gear in the direction that quickly disengages it from the flywheel.

STARTER CLUTCH SERVICE

Starter clutch problems usually are the result of worn parts or are caused by rust or dirt freezing up the parts. It is not usually a good idea to put grease or oil on the starter clutch parts because it may prevent good engagement and will collect gum and dirt. If the parts are clean, free of rust, and properly aligned, the clutch should engage firmly. Always inspect for wear and replace all badly worn parts. For example, replacing the notched starter cup without replacing worn, rounded starter dogs would soon cause the cup notches to wear out again. Replacement parts are usually inexpensive and ensure good operation. If several parts are badly worn, consider replacing the entire starter unit.

ELECTRIC STARTER—
BENDIX DRIVE

Figure 80

CHECK POINTS

1. Poor connection. Clean and tighten
2. Loose mounting bolts and stripped threads cause bad ground
3. Worn or dirty bendix drive allows starter to spin free
4. Worn or binding gear teeth
5. Worn bushings

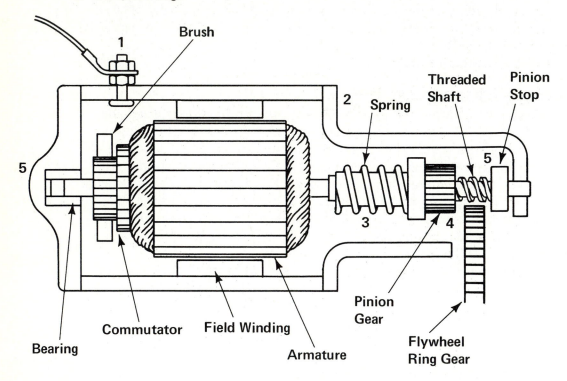

ELECTRIC STARTER—
BENDIX DRIVE

The electric starter is similar to the D.C. generator in that it has a field wound on field pole shoes and a turning armature that is connected to the circuit through the commutator and brushes. Generally, the armature winding contains much heavier wire to allow greater current flow and thus develop the great torque needed to turn the engine. Small gas engines do not pose the same problems as does the high-compression automotive engine. In addition to being much smaller, most small gas engines incorporate some type of compression-release technique to provide for easy pull rope starting. Speed reduction and torque increase are achieved by the very small starter pinion gear meshed with the large flywheel ring gear. Reductions of up to 20:1 are achieved.

The starter is most commonly engaged to the flywheel by a Bendix type drive that utilizes a spring loaded pinion gear fitted to a threaded shaft. The sudden start of the starter motor causes the pinion gear to thread itself out on the shaft until it hits the stop washer. Then the pinion gear must turn with the starter motor. As the pinion gear moves out on the shaft, it becomes meshed with the flywheel ring teeth. Now the starter motor is driving the flywheel through the pinion gear. When the engine starts, the spinning flywheel ring gear will "spin" the pinion gear back down the threaded shaft out of the way of the flywheel ring gear.

The starter is mounted securely to the engine by mounting flanges on either the side or end of the starter. Because the engine block may be made of cast aluminum and the size of the starter motor may be small, the mounting bolts could pose a problem. If a large wrench is used, the bolts may easily be overtightened and the threads stripped. Also, because of the vibration encountered, the bolts may become loose. If the bolts become loose, the starter will not make a good electrical connection to the engine ground or may allow the pinion gear to move away from the flywheel ring gear, causing improper meshing of the gear teeth.

Bendix drives may fail because of dirt, wear, or improper mounting. Dirty and worn Bendix drive units usually allow the starter motor to spin free without engaging the flywheel. The Bendix drive is easily replaced.

If the starter is not mounted firmly or not correctly aligned, the pinion gear may bind when meshing with the flywheel ring gear causing starter drag. This will cause slow starter operation or will cause the starter to be locked completely.

ELECTRIC STARTER TESTS

Figure 81

CAUTION: SAFETY GLASSES SHOULD BE WORN AND JEWELRY (RINGS AND
WATCHES) SHOULD BE REMOVED WHEN PERFORMING
ELECTRICAL TESTS!

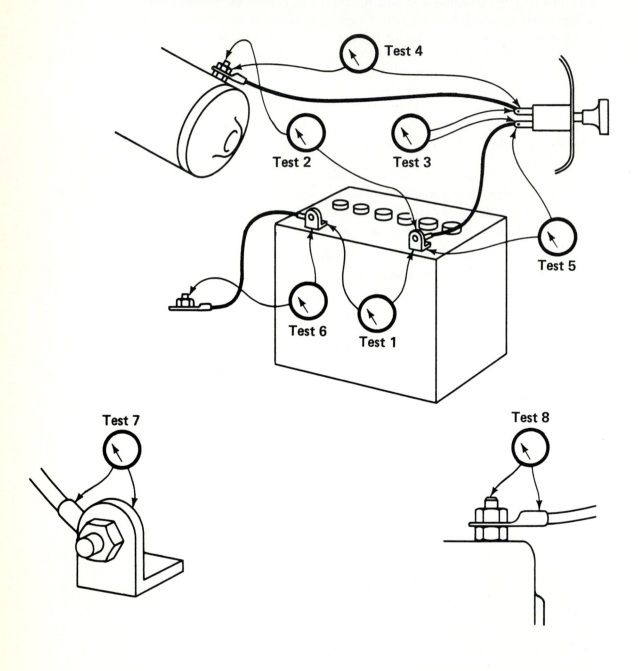

ELECTRIC STARTER TESTS

Starter circuits can be quickly checked with a D.C. voltmeter, for the voltmeter can be connected without disconnecting any circuit components. A series of voltmeter tests are identified in Figure 81.

First, test the battery voltage. Connect the voltmeter as shown in Test 1. Observe the battery voltage. It should be either 6.3 volts or 12.6 volts. Record the reading below. Push the starter switch and observe the battery voltage. Record the reading.

> TEST 1   Battery voltage with no load. _____
> Battery voltage with starter
> switch being held. _____

If the battery voltage remains the same or falls very little and the starter fails to crank the engine, either the switch, starter, or wiring is not making a complete circuit connection. Continue with voltmeter Tests 2, 3, 4, and 5 noted in Figure 81.

If the battery voltage falls below 4 volts for a 6-volt battery or 8 volts for a 12-volt battery, either the battery is not up or a short exists in the starter circuit. Test the battery. (See Battery Tests.)

> TEST 2   Meter reading with no load. _____
> Meter reading with starter
> switch being held. _____

Test 2 checks the switch and cables. Before cranking, the voltmeter will read the battery voltage, the same as Test 1. When the starter switch is engaged, the voltmeter should drop to zero or less than one-half volt. If the voltmeter does not drop down when the switch is engaged, a loss is occurring in the switch or cables.

> TEST 3   Voltage across switch before contact. _____
> Voltage across switch with switch
> being held. _____

To check the switch only, connect the voltmeter across the switch terminals as shown in Test 3. Again, the meter will read the battery voltage until the switch is engaged. When the switch is engaged, the meter should drop to near zero volts. Failure to drop to less than one-half volt indicates a bad switch.

If Test 2 showed a loss that Test 3 did not show, the loss occurred in a cable or connection. Tests 4, 5, and 6 (Figure 81) will show cable loss while cranking. Any reading indicates loss. Locate the loss and correct it. Corroded terminals or bad connections could cause loss. Test as shown in Tests 7 and 8 (Figure 81). These tests are made with the starter switch engaged. Any reading indicates loss.

STARTER-GENERATOR

Figure 82

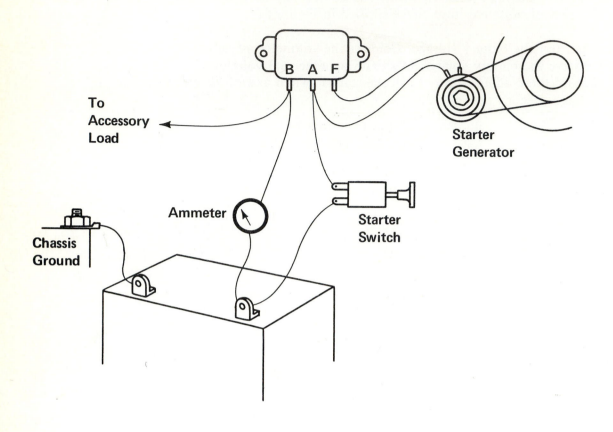

To
Accessory
Load

B A F

Starter
Generator

Ammeter

Starter
Switch

Chassis
Ground

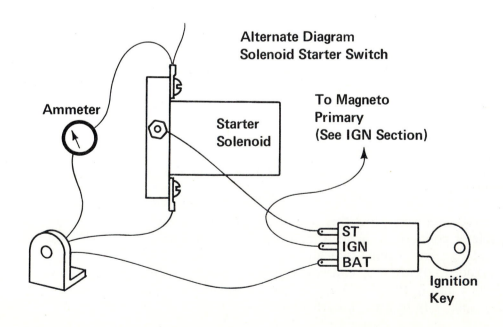

Alternate Diagram
Solenoid Starter Switch

Ammeter

Starter
Solenoid

To Magneto
Primary
(See IGN Section)

ST
IGN
BAT

Ignition
Key

STARTER-GENERATOR

The operation of the combination starter-generator refers to the purpose of the cutout relay in the regulator unit. When the output voltage of the generator becomes less than the voltage of the battery, the current stops flowing in the direction that charges the battery and it begins to flow back through the generator armature winding. The field and the armature are both now battery powered. The magnetic fields produced by the field windings and the armature are now like magnetic poles and are therefore repelling each other, causing the armature to attempt to turn to align the magnets. The generator is now acting as a *motor*. During generator operation the cutout relay opens when the generator output falls and it prevents the generator from "motoring."

A starter switch connected to bypass the cutout relay makes the generator become a starter motor when desired. The starter switch connects the armature directly to the battery positive terminal while the other armature brush is connected directly to the engine ground (battery negative). This makes both the field windings and the armature winding full strength for maximum starting torque.

The starter-generator is usually belt-driven to the engine. The engine pulley is larger to decrease the speed and increase turning torque while starting. The belt must be kept at correct tightness to crank the engine satisfactorily. Once the engine starts, the unit operates as a generator.

A solenoid switch and ignition key may be used instead of the heavy starter switch. Keep in mind here that, since a great deal of power is needed to crank the engine, heavy current flow is needed in the starter switch circuit. On the solenoid switch the starter switch is engaged electrically. The ignition key engages the starter switch.

Notice that the ignition key in the START position connects the solenoid coil to the battery positive for cranking. In the OFF position the key shorts the ignition primary winding to the engine ground and prevents the points from creating a spark and thus stops the engine. (See Ignition section.) *Caution:* This switch cannot be replaced with an automotive switch because an automotive switch connects the ignition terminal to the battery positive instead of to the ground. The magneto and switch would be burned out immediately.

The ammeter is also bypassed during the cranking operation and will not read the cranking current. The current needed for cranking is in excess of the capability of the meter and would cause meter damage. The charge from the regulator to the battery will show on the ammeter, as will flow from the battery to an accessory load. On some units there is an accessory terminal on the regulator.

Figure 83

CAUTION: DO <u>NOT</u> ATTEMPT TO CHECK OR ADJUST BELTS WITH ENGINE RUNNING!

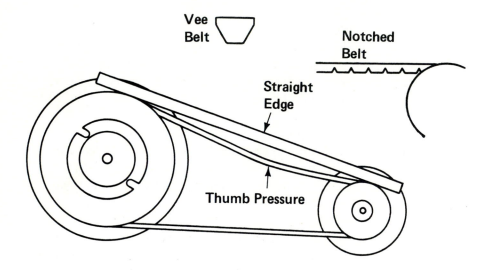

STARTER-GENERATOR TESTS

Starter-generator units are tested as separate units. The charging function is tested as any generator unit. On these units it is usually advisable to check the starter function first since the starter and generator are the same unit. If it will crank the engine, then it should also charge and if it will crank the engine but will not charge, the problem is likely in the external wiring or the regulator unit.

A frequent cause of failure of these systems is the drive belt. Check the pulleys to make sure that a smooth V-belt has not been placed on a unit designed for a notched belt. Check the belt tightness. It should not deflect more than one-quarter of an inch with normal thumb pressure. If the belt has been slipping and has become glazed, replace it. The new belt will need retightening after a few hours of operation and certainly after each operating season. If the unit is used two seasons such as mowing in summer and snow plowing in winter, the belt tightness should be checked *each* season.

VOLTMETER BATTERY TEST
Indication of starting or charging problems may be detected by connecting a D.C. voltmeter to the battery as shown on the next page. When connected to a battery on a unit not running, the voltmeter should read 12.6 or 6.3 volts, indicating a normal battery voltage. Observe the voltmeter carefully while cranking the engine. The meter should stay above 4 volts for a 6-volt battery and above 8 volts for a 12-volt battery. Once the engine is running, the voltmeter will indicate charging voltage. If the battery comes above the initial reading of 6.3 or 12.6 volts, the unit is charging. The amount of charge will be indicated on the ammeter if there is one. If there isn't an ammeter, one can be connected as shown by placing it in the battery cable at the regulator battery terminal. At half throttle, voltage readings should not exceed 7.5 volts or 15 volts.

STARTER-GENERATOR TESTS
(cont.)

Figure 83 (continued)

CAUTION: SAFETY GLASSES SHOULD BE WORN AND JEWELRY (RINGS AND WATCHES) SHOULD BE REMOVED WHEN PERFORMING ELECTRICAL TESTS!

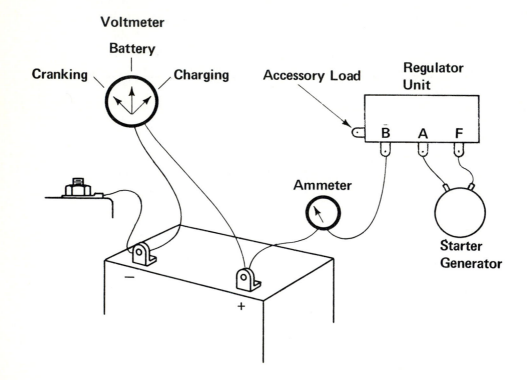

STARTER-GENERATOR TESTS

PROBLEM	CAUSE
Starter spins but does not turn engine.	Loose belt. Glazed belt. Wrong type belt.
Battery does not stay fully charged.	Loose belt.
	Recharge battery. If starter functions OK, fault is in wiring or regulator. Test as for any generator.
Battery stays fully charged but will not crank or cranks slowly.	Problem in the starter switch circuit. Check starter switch and wiring as described in Starter Test section.
Voltage drops below 4 volts on a 6-volt system or 8 volts on 12-volt system when cranking.	Battery defective or not fully charged. Starter shorted.
Battery voltage remains unchanged when starter switch is pushed. Charging circuit OK.	Defective starter switch, loose or corroded connection, defective ground on battery or starter.
Battery voltage drops too low when cranked. Voltage does not rise above battery normal after engine starts.	Battery discharged because of no charging circuit output. Check charging circuit.

T F 1. Chemical additives are needed to prolong the life of a storage battery. Page 203.

T F 2. Battery life can be restored by dumping out the old water and adding new distilled water. Page 203.

T F 3. A battery gives off an explosive gas while being charged. Page 203.

T F 4. Always turn a battery charger off before disconnecting the cables. Page 203.

T F 5. Dirt buildup on the outside of a battery is not harmful if it does not enter the battery. Page 205.

T F 6. Always wear safety glasses when working with a battery. Page 204.

T F 7. If at least five of the six cells read fully charged, the battery is considered good. Page 205.

T F 8. A nearly dead battery will read full voltage under no load. Page 207.

T F 9. Rewind starter problems are caused by a lack of oil on moving parts. Page 213.

T F 10. Some small gas engines use the generator as a starter motor. Page 219.

T F 11. If the ignition switch fails, any good automotive switch can be used. Page 219.

T F 12. If an ammeter is included on the unit, it will indicate starter current draw. Page 219.

T F 13. The electrical charge is chemically stored in the battery. Page 203.

T F 14. The electrolyte solution in the battery changes as the battery is charged or discharged. Page 203.

T F 15. All batteries have a vented cap for each cell. Page 207.

T F 16. Repair of a worn, mechanical-type rewind starter usually involves replacement of worn parts. Page 211.

T F 17. The electric starter is commonly engaged to the flywheel gear by a dog clutch. Page 215.

T F 18. Bendix type starter drives may fail because of dirt or wear. Page 215.

T F 19. Before testing the starter, one must test the battery voltage. Page 217.

20. List two causes of water loss from a battery that is in use. Page 203.

21. Describe the voltmeter test for self-discharge. Page 205.

22. What is an acceptable minimum voltage the battery should be able to maintain while cranking the engine? Page 207.

23. List the common causes of electric starter failure. Page 223.

24. Identify problems that may occur if the battery is not securely mounted. Page 203.

25. What causes a battery to self-discharge? Page 205.

26. At 80°F, what hydrometer reading will be obtained from a fully charged battery? Page 205.

27. How can a "sealed" battery be tested? Page 207.

28. How can one test the battery capacity? Page 207.

29. What is the condition of the battery if the voltage remained above 9.5 while the engine was cranked? Page 209.

30. Explain how one should clean the battery. Page 205.

CHARGING SYSTEM FUNDAMENTALS AND ALTERNATOR TESTING

Figure 84

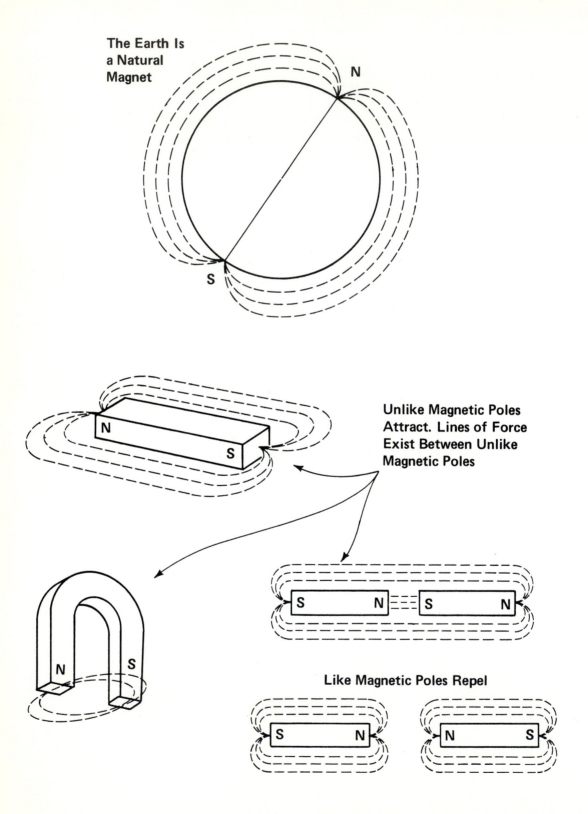

The Earth Is
a Natural
Magnet

Unlike Magnetic Poles
Attract. Lines of Force
Exist Between Unlike
Magnetic Poles

Like Magnetic Poles Repel

MAGNETS

There are three basic types of magnets: natural magnets, permanent magnets, and electromagnets. The magnetic fields set up by these magnets are alike, that is, they each have a north pole and a south pole. They differ only in the way in which the magnetic field is produced and retained. Because unlike magnetic poles attract, the south pole of a permanent magnet is attracted to the north pole of a D.C. electromagnet just as it would be to the north pole of another permanent magnet.

NATURAL MAGNETS

The Earth is itself a natural magnet because it has a definite north and south magnetic pole. Some ores are also natural magnets. Because they are not strong enough or durable enough for use on small gasoline engines, natural magnets will not be discussed further.

PERMANENT MAGNETS

Permanent magnets are not really "permanent." Permanent magnets are made by humans and can be destroyed at any time. Hardened steel can be magnetized and will retain its magnetism for a long period if it is handled properly. Special alloy magnets now available have extremely long lives.

A screwdriver is a good example. The hardened blade of a screwdriver may become magnetized and will remain magnetized for many years. Some screwdrivers are purposely magnetized to help hold screws or to pick up objects. A screwdriver so magnetized is a permanent magnet.

Good permanent magnets are usually in the form of a bar or horseshoe. Like natural magnets, they have a definite north and south magnetic pole. A strong attraction exists between the north and south magnetic poles that causes lines of force to be set up between the poles. Study the diagrams on lines of force in the illustration on the opposite page.

DESTROYING THE PERMANENT MAGNET

Magnetism in a hardened steel permanent magnet is a result of molecular alignment. The molecules may be returned to their normal random arrangement and thus destroy the magnetism. Permanent magnets such as flywheel magnets may be destroyed by a sharp blow; for example, by dropping, heating, or by exposing them to alternating current magnetic fields such as large electric motors.

ELECTROMAGNETS

Figure 85

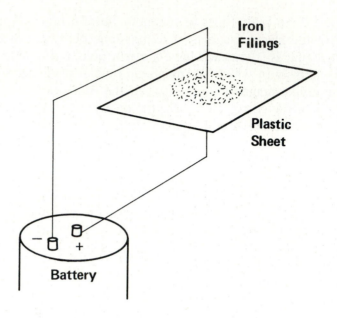

Iron Filings

Plastic Sheet

Battery

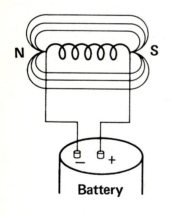

Magnet Has Combined Strength of All Turns

Battery

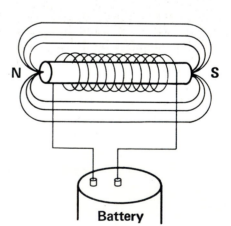

Soft Iron Bar Strengthens and Concentrates the Magnetic Field

Battery

ELECTROMAGNETS

When electric current is passed through a wire, a magnetic field surrounds the wire. The illustration at the left shows how soft iron filings will be attracted in a circular pattern to a wire carrying current. A coil of wire made up to several turns wound closely and continuously will produce a strong magnetic field. The magnetic field produced by each turn of wire adds to the magnetic field produced by all the other turns of wire to form a single strong field around the coil. As with the natural magnet and the permanent magnet, the electromagnet will have a north and south pole. Reversing the battery terminals will reverse the current flow through the coil and will reverse the north and south poles.

If the coil is wound on a soft iron core, the magnetic field will become concentrated and much stronger. Most electromagnetic devices used on engines (the ignition coil, generator, and alternator) have coils of wire wound on laminated steel to improve the magnetic strength. The laminated layers of steel are insulated from each other to prevent currents through the core that would decrease the magnetic strength and produce heat. Sometimes rust will bridge across these laminations and will seriously cut the efficiency of the electromagnet.

The strength of the magnetic field surrounding a wire is dependent on the amount of current flowing through it. The strength of the magnetic field around a coil of wire depends on:

1. *The size of wire used.* The larger the wire size, the easier it is for the current to flow. Thus, more current flow and a stronger magnet are produced.
2. *The number of turns of wire.* Because the magnetic field of each turn of wire adds to that of the adjacent turns, more turns of wire will produce a stronger magnet.
3. *Increase of the voltage to the coil.* Increasing the electrical pressure (voltage) will cause more current to flow. If the voltage is increased to the point that the current flow exceeds the wire size, the coil will heat up, causing damage to the coil.

RESIDUAL MAGNETISM

Figure 86

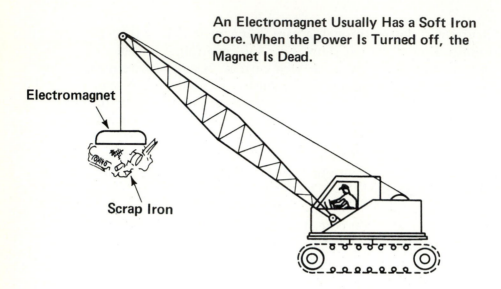

An Electromagnet Usually Has a Soft Iron Core. When the Power Is Turned off, the Magnet Is Dead.

Electromagnet

Scrap Iron

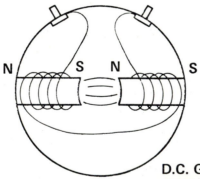

N   S   N   S

D.C. Generators Are Designed So That Residual Magnetism Remains After the Field Coils Are Turned off. This Makes the Generator Self-exciting. It Can Start Charging Without Help from the Battery.

RESIDUAL MAGNETISM

Note that in making an electromagnet a soft iron core was used. A soft iron core will act to concentrate the magnetic fields satisfactorily, but it will lose its magnetic characteristics as soon as the current stops flowing. It is usually desirable to have an electromagnet that will shut off completely when the current flow is stopped; for example, one used to load scrap iron. If the coil is wound on a core that has been hardened, the hardened core will retain some of the magnetic strength of the electromagnet. This remaining magnetism is referred to as *residual magnetism*.

If the core is sufficiently hardened, it will become a permanent magnet, that is, it will retain considerable magnetic strength after the current flow in the winding has been stopped. If not destroyed, this remaining magnetism could remain for several years.

The pole shoes of a D.C. generator retain some of the magnetic strength of the field coils around them. This makes the generator self-exciting and makes it possible for the generator to generate when it is not connected to a battery. An automotive *alternator* usually does not have this characteristic. It will not charge if the battery is completely dead because it relies on the battery to supply the excitation or magnetic field. If the battery is too weak to create a magnetic field in the alternator, the alternator will not charge.

CREATING CURRENT FLOW

Figure 87

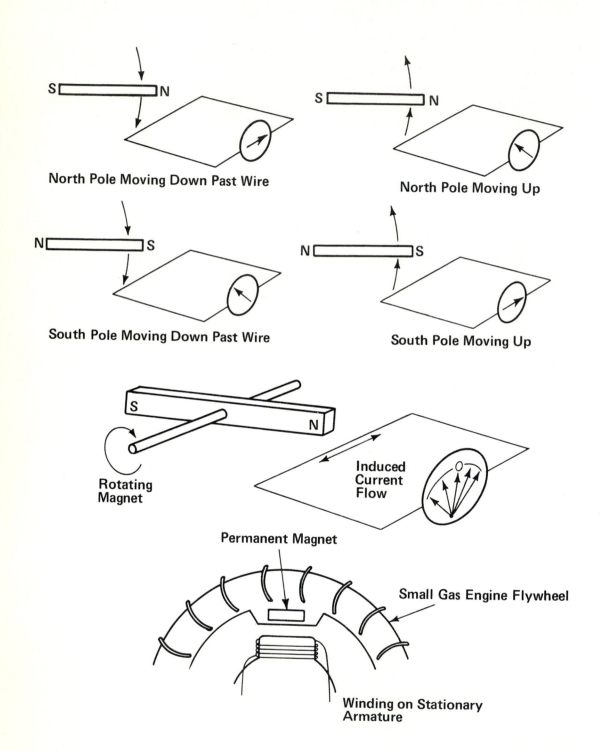

North Pole Moving Down Past Wire

North Pole Moving Up

South Pole Moving Down Past Wire

South Pole Moving Up

Rotating Magnet

Induced Current Flow

Permanent Magnet

Small Gas Engine Flywheel

Winding on Stationary Armature

CREATING CURRENT FLOW

Current flow was induced in the primary winding of the ignition coil by a moving magnet. The same principle is utilized in the charging system whether it is the D.C. or A.C. type. As you recall, moving a magnet rapidly past a conductor will induce current flow. In the upper illustration on the opposite page, the magnet is moved rapidly past the conductor and the current flow will be indicated on the ammeter. If the magnet is moved back past the conductor in the opposite direction, the current will flow back the other way as indicated by the direction of deflection of the ammeter.

Moving a south pole past the wire will create current flow in one direction. Moving a north pole past the wire will cause current flow in the opposite direction.

PRODUCING ALTERNATING CURRENT
If the magnet were attached to a shaft and rotated, current flow would be induced in the wire first in one direction when the north pole moves past and would be induced in the other direction as the south pole moves past. This back and forth movement of current would cause the ammeter to deflect both directions from the center zero point. If the magnet is turned fast, the meter would not be able to deflect fast enough and would not accurately read the amount of current flow. Damage to the meter could result. This continuous back and forth movement of current is called ALTERNATING CURRENT (A.C.). The current produced by a battery that is always flowing the same direction is called DIRECT CURRENT (D.C.).

PRODUCING A STRONGER CURRENT
The current produced by moving the magnet past one wire would be very small. Increasing the wire area being passed by the magnet will increase the current flow. If the wire is wrapped into a coil, the current will be induced in each turn of the coil. The current induced in each turn of the wire adds to the current of all the turns producing a strong current flow. Winding the coil of wire on a laminated iron core will help the passing magnet set up the magnetic lines of force through the winding, which will increase the current flow in the winding.

Typical small gas engine alternators use a series of coils (stator winding) mounted on a laminated ring (stator) under the flywheel using the flywheel magnet to induce the current flow in the stator winding. The ignition coil may or may not be integrated into the same set of coils.

Keep in mind that this is alternating current at this point and is not suitable for charging the battery because the current flow is equal in each direction and no charging of the battery would occur.

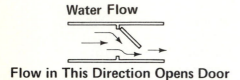

Water Flow

Flow in This Direction Opens Door

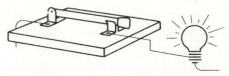

A Diode Acts as a Closed Switch and Allows Current to Flow Freely in One Direction

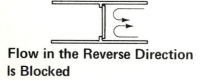

Flow in the Reverse Direction Is Blocked

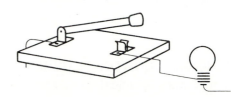

But Acts as an Open Switch and Prevents Flow in the Reverse Direction

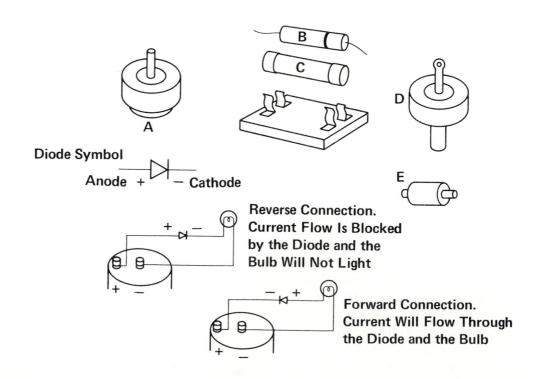

Diode Symbol

Anode + — Cathode

Reverse Connection. Current Flow Is Blocked by the Diode and the Bulb Will Not Light

Forward Connection. Current Will Flow Through the Diode and the Bulb

DIODES

A *diode* is a one-way electrical check valve. It permits the electrical current to flow easily in one direction, but it stops the current flow in the other direction. Just as a check valve will permit water to flow in one direction and will block flow in the opposite direction, a diode performs this function in an electrical circuit.

One use for the diode in the electronic ignition system is to allow current to flow freely from the charging coil to the capacitor, thus charging the capacitor. The diode prevents the current from flowing back through the coil trapping the charge on the capacitor. This is a typical use for a diode.

The diode might also be compared to an electrical switch that is quickly closed so that the current can flow in one direction, but it is quickly opened to prevent current flow in the opposite direction.

Diodes are available in a number of shapes and sizes, depending on the job they are performing. Diodes are rated by current capacity and maximum voltage.

The CURRENT rating is the maximum amperes the diode can pass in the forward direction. Current flow in excess of this amount will blow the diode.

The VOLTAGE rating determines the maximum voltage that can be applied to the diode in the reverse direction when no current is flowing without forcing its way across or bridging the diode. Remember that voltage is electrical pressure and that high voltage is capable of jumping distance. High-voltage diodes are very different from those used in low-voltage circuits. High-voltage diodes usually look like illustration B on the opposite page.

A HEAT SINK is necessary for high-current diodes such as shown in illustrations A and D on the opposite page. The type shown in A is used on larger alternators. It is pressed into the alternator frame or isolated heat sink. The type shown in D is bolted to a heat-conducting surface. In both cases the heat sink becomes part of the circuit because there is only one connecting terminal on the diode.

Most diodes used with alternators on small gas engines are like C and E on the opposite page. They do not carry enough current to need a heat sink, and they usually fit into a clip connection for easy removal.

DIODES (cont.)

Figure 88 (cont.)

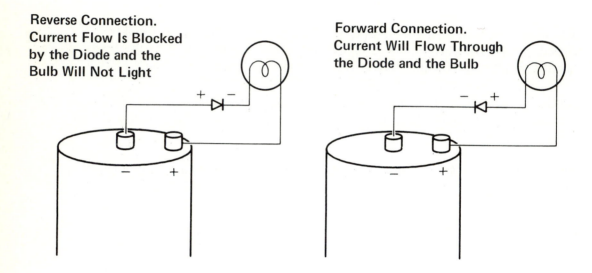

Reverse Connection.
Current Flow Is Blocked
by the Diode and the
Bulb Will Not Light

Forward Connection.
Current Will Flow Through
the Diode and the Bulb

DIODE TESTING

Diodes may be tested with an ohmmeter. Connecting the lead to the diode one way should show low resistance. Connecting the lead the other way should show high resistance. If the diode shows either high or low resistance both ways, it is defective.

The circuit depicted in the accompanying illustration showing the dry cell and light bulb connector through a diode may be used as an effective diode tester. Place the diode in the circuit both ways. The bulb should light with the diode connected one direction and should not light at all if connected the other way. If the bulb lights both ways, the diode is shorted. If the bulb will not light either way but will light when connected without the diode, the diode is open. If the light glows dimly during the reverse CON-NECTION, the diode is leaking and should be replaced.

Figure 89

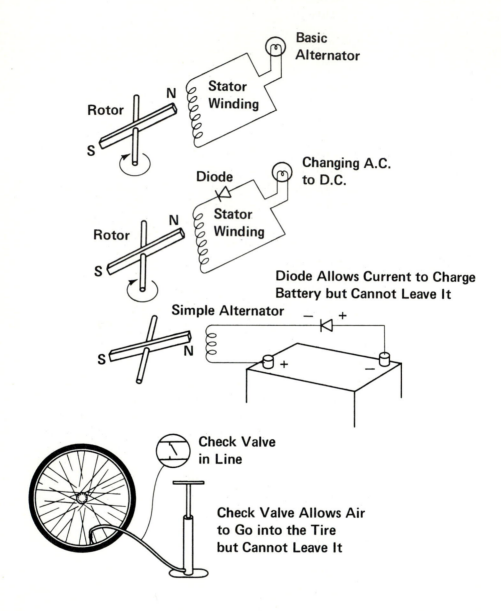

Basic
Alternator

Stator
Winding

N

Rotor

S

Changing A.C.
to D.C.

Diode

Stator
Winding

N

Rotor

S

Diode Allows Current to Charge
Battery but Cannot Leave It

Simple Alternator

S

N

− +

+

−

Check Valve
in Line

Check Valve Allows Air
to Go into the Tire
but Cannot Leave It

THE BASIC ALTERNATOR

Study the diagrams on the opposite page carefully. In the upper diagram we have the basic alternator, a stationary (stator) coil, and a rotating magnet (rotor). Remember that as the rotor turns, the alternating north and south poles will cause the current to flow back and forth in the circuit (alternating current). The light bulb does not care that this is alternating current (A.C.) because the A.C. will heat the filament and produce light just as will direct current (D.C.).

DIODE RECTIFIED OUTPUT

In the next diagram a diode has been placed in series with the circuit so that all current must flow through the diode. As one magnetic pole passes the stator winding, it will produce current flow in the circuit if the current flow is in the forward direction of the diode. As the other magnetic pole passes the stator, it will try to induce current flow in the reverse direction to which the diode acts as an open switch allowing no current flow. The circuit is shut off by the diode.

Now the bulb will be lighted by pulses of current all moving in the same direction. When the current flow is all in one direction, it is called direct current (D.C.). The bulb is now operated on pulses of D.C. In this circuit A.C. has been changed to pulsating D.C.; this is called *rectifying*. The diode is one type of rectifier that will change A.C. to D.C.

CHARGING THE BATTERY

When this circuit is connected to a battery, the pulses of D.C. can be used to charge the battery. Notice that the diode is placed in the circuit in reverse to battery polarity. The battery cannot discharge through the diode and coil. The diode is placed so that the current can flow in the direction that charges the battery but cannot flow in the direction that will discharge the battery.

Compare the illustration of the tire pump with the simple alternator above it. When the pump pressure becomes greater than the tire pressure, the air moves through the check valve into the tire. As the pump moves back up, the air is prevented from coming back out of the tire by the check valve (enlarged). The back and forth movement of the pump is changed into a one-way movement of air by the check valve.

When the voltage in the stator becomes greater than the voltage of the battery, the current will flow, charging the battery. When the current tries to flow the other direction in the stator, which would discharge the battery, the diode acts as an open switch and prevents the current flow.

LOW-CURRENT
ALTERNATORS

Figure 90

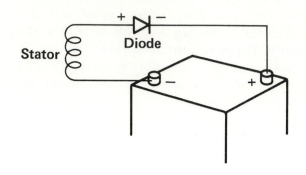

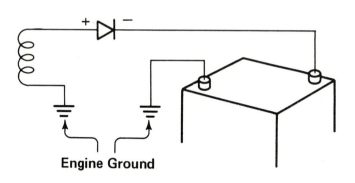

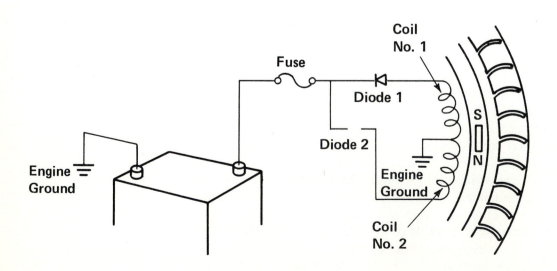

LOW-CURRENT ALTERNATORS

Typical charging systems use the engine as one side of the electrical circuit. This is called the *grounded* side. Compare the two circuit diagrams in the illustration opposite. On both diagrams, a complete circuit exists from the battery negative (−) terminal to the stator coil, to the diode, and to the battery positive (+) terminal. Using the engine as one side of the circuit eliminates some of the wiring and reduces confusion for there are fewer wires to keep straight and less danger of hooking something up backward as the engine block is always the same as the battery negative terminal. On some older cars and tractors, the battery was reversed and the engine (entire chassis) became the same as the battery positive terminal. They were referred to as *positive ground* systems. It is important to check this out before removing the battery cables or attempting any tests on the circuit.

The low-current alternator diagram shows two charging coils mounted under the flywheel. The flywheel magnet, or magnets, induces current flow in each of the coils as it passes. Note that each coil has one lead connected to the engine ground, which in turn is connected to the battery negative terminal. The other lead from each coil is connected through its diode to a common fuse and to the battery positive terminal.

As the flywheel magnet passes charging coil No. 1, the current flow induced one direction in the coil is passed through diode No. 1 to charge the battery. The current flow induced in the opposite direction is blocked by the diode. The magnet continues on, inducing A.C. current in coil No. 2. Diode No. 2 allows current flow to charge the battery and blocks the reverse flow.

This is an alternator typically used on lawn and garden tractors. It is simple, inexpensive, and trouble free if it is not abused.

Because charging coils are quite small, they limit the output of the system. And because the output is limited, it is not likely that the battery will become overcharged during use; thus, no voltage regulator is supplied with this system. If the engine is operated for long periods of time without restarting, the battery will be overcharged by the system. The first symptoms of overcharging will usually be excessive water consumption by the battery. Removing one of the diodes will cut the charging system output and will prevent overcharging during extended engine operation. If the engine is used only for short periods calling for frequent starter use, an additional battery charge may be necessary occasionally.

Because the alternator is an A.C. current device, it is self-current regulating and no current regulating device is needed as with D.C. generators. Because the diodes prevent the return of the battery current to the alternator when the alternator is not in use, a cutout relay is not needed.

FLYWHEEL ALTERNATOR SYSTEM TESTING

Figure 91

CAUTION: SAFETY GLASSES SHOULD BE WORN AND JEWELRY (RINGS AND WATCHES) SHOULD BE REMOVED WHEN PERFORMING ELECTRICAL TESTS!

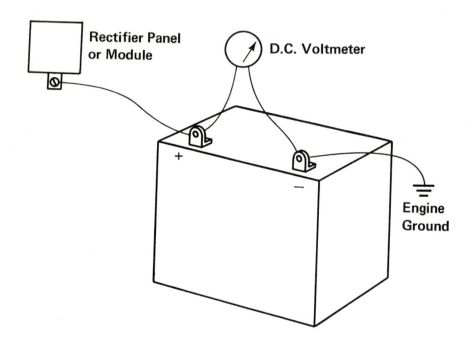

FLYWHEEL ALTERNATOR SYSTEM TESTING

Although flywheel alternator systems vary somewhat, they are similar in operation and lend themselves to some common tests. Many of the components have been molded into modules, which if found defective by simple tests must be replaced.

OUTPUT VOLTAGE TEST

Before starting the engine, connect a D.C. voltmeter that will read at least 16 volts to the battery terminals. Read the battery voltage carefully and record.

Battery volts with engine stopped (should be 12.6 volts)

_____

Crank the engine briefly. Recheck the battery volts.
Battery volts after short load (should be 12.6 volts)

_____

Start the engine. Allow the engine to run at 3,000–3,600 rpm near wide open for one minute. CAUTION: Do not rev an unloaded engine. Read the battery voltage while the engine is still running 3,000–3,600 rpm.

Battery voltage, engine running rated rpm

_____

TEST RESULTS	PROBLEM
Battery volts before running are 12.6 volts or higher. If higher, probably the battery has recently been charged.	OK.
Battery volts are less than 12.6.	Check battery. See Battery Test section.
While the engine is running, the battery volts are higher than before running.	Charging circuit OK.
Battery volts do not increase after engine has been run 3,600 rpm for one minute.	Possible charging circuit problem.
Battery volts during running are 15 volts or higher.	Overcharging. If it is an unregulated unit, remove one diode during extended operation.
Battery volts increase, but battery becomes discharged periodically.	Reduced output. Check diodes and perform ammeter test. Possibly the unit is being started frequently. Boost charge battery periodically.

Figure 91 (cont.)

CAUTION: SAFETY GLASSES SHOULD BE WORN AND JEWELRY (RINGS AND WATCHES) SHOULD BE REMOVED WHEN PERFORMING ELECTRICAL TESTS!

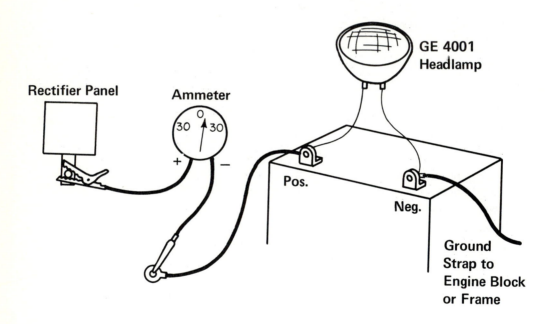

FLYWHEEL ALTERNATOR SYSTEM TESTING (cont.)

AMMETER OUTPUT TEST

An ammeter can be used to determine the amount of charge being placed on the battery. Connect a D.C. ammeter (available at automotive stores) in series with the charging lead. Remove the charging lead from the rectifier. Clip one lead from the ammeter to the rectifier output and the other lead to the battery charged lead. The output current will now flow through the ammeter. Bring the engine speed to 3,600 rpm and observe the ammeter reading. Look carefully. A very low output may be difficult to detect on a 30-ampere meter.

If little or no output is observed, use an automotive headlamp (GE No. 4001) as a battery load. Connect the lamp directly to the battery terminals with test leads. Again, bring the engine to 3,600 rpm and read the ammeter carefully. Failure to produce output indicates charging circuit problem. Continue the charging circuit tests.

FLYWHEEL ALTERNATOR COMPONENT TESTING

Figure 92

CAUTION: SAFETY GLASSES SHOULD BE WORN AND JEWELRY (RINGS AND WATCHES) SHOULD BE REMOVED WHEN PERFORMING ELECTRICAL TESTS!

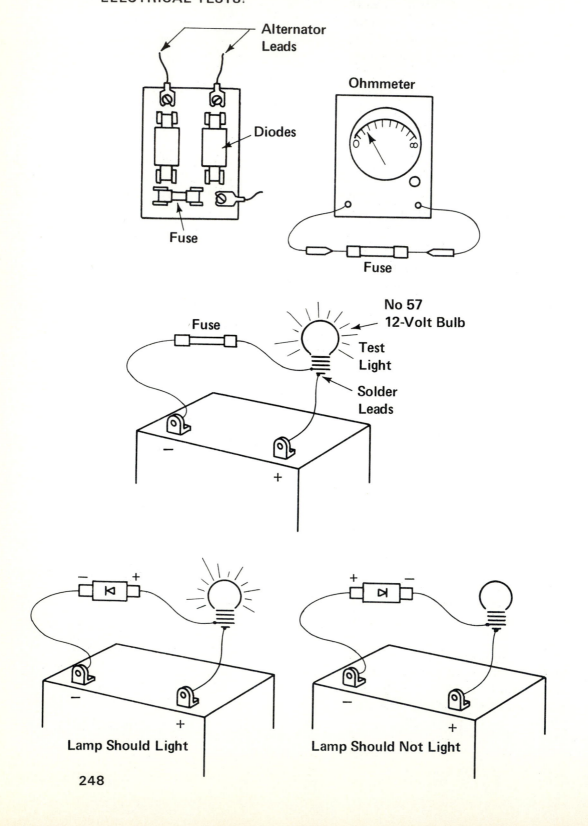

FLYWHEEL ALTERNATOR COMPONENT TESTING

As you have seen on the diagrams of the simple flywheel alternator, there are no friction-causing moving parts and, unless abused, they usually cause no trouble. Components such as the fuse and the diodes will take very little abuse. A loose wire or terminal can make momentary opens or shorts in the circuit that could blow the fuse or diodes. A visual check of leads, plugs, terminals, and the clip-in mountings may reveal such a problem.

To check the fuse, remove it from the fuse holder and test for continuity with an ohmmeter. If an ohmmeter is not available, a small light bulb and battery can be used. A good continuity tester can be made by soldering test leads to a bulb. Placing the fuse in question in series with the bulb and battery so that the current must flow through the fuse to light the bulb will determine if the fuse is still good (continuous). Clean the fuse and fuse holder terminals before replacing the fuse. Use emery paper or fine sandpaper if necessary to assure good contact.

Check the diodes with an ohmmeter or battery and test light as shown. In one direction, the diode should read ZERO ohms. Reverse the diode and the meter should read infinity or high resistance. If the diode is connected in series with a test lamp and battery, the lamp should light when connected one direction and should not light when reversed. If the diode symbol is printed on the diode, or if the polarity is marked, the lamp should work as shown in the bottom two diagrams in the accompanying illustration.

DIODE TEST RESULTS	CAUSE
Continuity one direction. No continuity when reversed.	Good diode.
Continuity both directions.	Shorted diode. Replace.
No continuity either direction.	Open diode. Replace.
Continuity one direction. Some continuity when reversed. Low meter reading or lamp glow.	Leaking diode. Replace

Clean the diode contacts the same way as the fuse contacts. Check all wires and connections. When replacing covers make sure that wires are not pinched, thus causing a short circuit.

To check the stator winding, use the ohmmeter or test lamp and battery to check continuity through the coil winding. A break in a wire or a bad connection will be detected by this test. Disconnect the alternator leads from the diodes before testing. Test each coil.

REGULATED FLYWHEEL ALTERNATOR

Figure 93

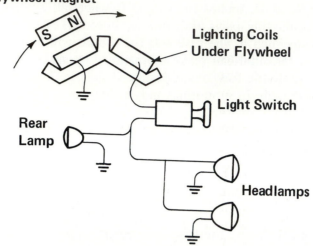

Flywheel Magnet

Lighting Coils Under Flywheel

Light Switch

Rear Lamp

Headlamps

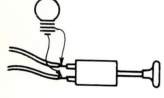

Test Lamp

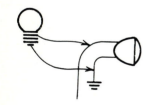

CAUTION: SAFETY GLASSES SHOULD BE WORN AND JEWELRY (RINGS AND WATCHES) SHOULD BE REMOVED WHEN PERFORMING ELECTRICAL TESTS!

1. Test the Switch. If the Lamp Lights When the Switch Is ON, Switch Is Defective.

2. Test the Bulb. If the Test Lamp Lights, Headlamp Is Defective.

3. Check the Ground. Connect Test Lamp to Headlamp Ground Terminal and to a Good Ground. If Test Lamp Lights, Ground Was Bad.

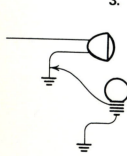

4. Test Stator Output.

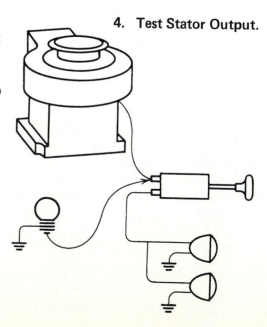

REGULATED FLYWHEEL ALTERNATOR

The unregulated flywheel alternator will do an acceptable job of charging the battery on lawn and garden equipment that is normally run for a predictable time between each start. As pointed out earlier, if an unregulated unit is operated for short periods or at low speeds, external recharging may be necessary occasionally to supplement the small output of the flywheel alternator. If the unregulated unit is operated for long periods at high rpm, the battery will be overcharged. The REGULATED flywheel alternator overcomes these disadvantages with increased output to recharge the battery faster and a voltage regulator to prevent overcharging.

The REGULATOR COIL is added to the charging coils. One lead from the regulator coil is connected to the battery (+) plus lead of one of the A.C. coils. The other lead goes to ground through the SCR. The SCR, when electrically switched ON, allows current flow in the regulator coil. The current flow in the regulator coil sets up a magnetic field that opposes the A.C. coils reducing their output. When the SCR is not ON, there is no current flow in the regulator coil and it has no effect on the charging coils, permitting full output.

The SILICON CONTROLLED RECTIFIER (SCR) is an electronic switch that has no moving parts. When a small voltage is applied to the gate connection, the SCR is turned ON. Until voltage appears at the gate, the SCR remains OFF to current flow through the SCR from the regulator coil to ground.

A ZENER DIODE is a very sensitive electronic device that allows a small amount of current flow only when the applied voltage exceeds a certain level (breakdown voltage). Zener diodes are available in a wide range of voltages. When the voltage reaches the firing voltage of the Zener diode, the diode turns ON (fires), allowing current flow. In this circuit the Zener diode is connected to battery positive. As the battery becomes charged, the voltage begins to rise above the normal battery voltage (12.6 volts). The Zener diode is usually set to fire at about 14.5 volts. When the battery voltage reaches 14.5 volts, the Zener diode fires, applying voltage to the gate of the SCR. The SCR is turned ON permitting current flow through the regulator coil, which opposes the A.C. coils and thus reduces output.

The VARIABLE RESISTOR may be used to permit an adjustment in regulator output. Changing the resistor would change the firing voltage of the Zener.

The CAPACITOR is quickly charged by the current flow through the Zener diode to build up the voltage to cause the gate to fire the SCR.

The SCR, Zener diode, resistor, and capacitor are all molded in a compact MODULE and are not accessible for testing. If an overcharging or under-charging condition exists, check the battery condition (see Battery Tests), check the fuse and diodes (see Component Tests), and check physically all wires and connections. If these are not at fault, replace the regulator module or check the manufacturer's literature for specific checks.

Figure 94

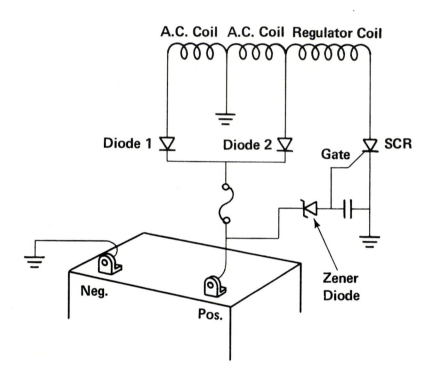

A.C. LIGHTING CIRCUIT

Some lawn and garden tractors are equipped with a headlamp that is powered directly from A.C. charging coils. The lights are not connected to the battery in any way. Some of these systems appear on units without battery or electric start or they may be used with a flywheel alternator and electronic ignition. The A.C. lighting system is completely independent of the battery charging circuit. The charging coils are sometimes contained in the same molded module.

The A.C. lighting circuit is similar to the low output flywheel alternator except that it does not require the rectifying diodes. The lights are operated with the alternating current created in the charging coils. The brightness of the headlamp depends on engine speed. This circuit has been common on motorcycles for several years.

To test the circuit, use a 12-volt test light with lead attached as for diode testing. Operate the engine at about one-half throttle with the light switch ON and make these tests.

1. Test the switch. Connect the test lamp to the switch terminals. If the lamp lights, the switch is defective.
2. Test the bulb. Connect the test lamp to the bulb connections. If the test lamp lights, the bulb is defective.
3. Test for headlamp ground. Connect the test lamp to the headlamp ground connections and to a good engine ground. If the test lamp lights, the headlamp is not making a good ground. Inspect carefully for bad connections and *rust*. Clean all connections.
4. Test for stator output. Turn the light switch OFF. Connect the test lamp to the charging coil lead at the light switch. Touch the other test lamp lead to a good engine ground. If test lamp lights, the stator is producing current and the problem is one of the above. Check wiring carefully for bad connections, frayed insulation, loose connections, or rust.

Questions for Section XI

T F 1. Electromagnets are usually wound on laminated layers of soft iron to reduce heat. Page 231.

T F 2. If rust bridges across the laminations of an electromagnet, its efficiency may be greatly reduced. Page 231.

T F 3. An ammeter must be connected *in* the circuit so that current flows through the meter. Page 247.

T F 4. Alternating current cannot be used to charge a battery. Page 235.

T F 5. Diodes are rated according to their maximum current and maximum temperature. Page 237.

T F 6. A diode is like an electrical switch that closes to pass current freely in one direction and opens to stop current flow in the other direction. Page 237.

T F 7. Some diodes require a heat sink. Page 237.

T F 8. A diode can change A.C. current to pulsating D.C. Page 241.

T F 9. When the engine is used as electrical ground, current will actually pass through the engine block. Page 243.

T F 10. Under no circumstance should the positive battery terminal be connected to ground. Page 243.

T F 11. Flywheel alternators may not have a current regulator or a voltage regulator. Page 243.

T F 12. If the battery loses water on a flywheel alternator system, removal of one diode may correct the problem. Page 243.

T F 13. Flywheel alternators require a cutout relay. Page 243.

T F 14. The power source for the A.C. lighting system is similar to the flywheel alternator without the rectifying diodes. Page 253.

15. Describe the three basic types of magnets. Page 229.

16. List three ways a permanent magnet can be destroyed. Page 229.

17. Name three ways the strength of an electromagnet can be increased. Page 231.

18. What is meant by *residual magnetism?* How is it achieved? What is a practical application? Page 233.

19. Draw or build a diode tester. Connect the diode both ways and explain the results. Page 239.

INDEX

V

Valve adjust, 165
Valve guides, 149
Valve install, 165
Valve lifters, 149
Valve removal, 125
Valve seats, 151
Valve service, 153

Venturi, 89
Vibration, 15
Voltage, 53

W

Winter storage, 17